SUNSET AT SEA GROVE, 1776.

SCHEYICHBI AND THE STRAND,

OR

EARLY DAYS ALONG THE DELAWARE.

WITH

AN ACCOUNT OF RECENT EVENTS AT SEA GROVE.

CONTAINING

SKETCHES OF THE ROMANTIC ADVENTURES OF THE PIONEER COLONISTS; THE WONDERFUL ORIGIN OF AMERICAN SOCIETY AND CIVILIZATION; THE REMARKABLE COURSE OF POLITICAL PROGRESS AND MATERIAL IMPROVEMENT IN THE UNITED STATES, AS SHOWN IN THE HISTORY OF NEW JERSEY, WITH PROOF OF THE SAFETY AND BENEFIT OF DEMOCRATIC INSTITUTIONS, AND THE NECESSITY OF RELIGIOUS FREEDOM.

TO WHICH IS APPENDED A GEOLOGICAL DESCRIPTION OF THE SHORE OF NEW JERSEY.

BY

EDWARD S. WHEELER.

ILLUSTRATED WITH TWELVE FULL-PAGE ENGRAVINGS, FROM ORIGINAL DRAWINGS BY D. B. GULICK, CHARLES W. KNAPP, AND OTHERS.

PRESS OF J. B. LIPPINCOTT & CO.
PHILADELPHIA:
1876.

DEDICATION.

TO MY CHRISTIAN FRIENDS.

who, firm in the faith themselves, can nevertheless respect the convictions of others; to earnest Christians whose spiritual trust and faith is so perfect, they have no fear any fact can disprove truth, or human error annul the divine law; to Christians whose character honors their creed, whose fairness and honesty command regard, while their kindness and courtesy inspire fraternal love; to all who love truth better than their own conceit; to all who reverence God more than any theory; to all who seek the good, the true, and beautiful themselves, and devoutly labor for the welfare and eternal happiness of humanity, I dedicate this volume.

ERRATA AND CORRIGENDA.

On page 5, 26th line, for "its citizens discovered" read: their citizens discovered.

On page 8, 32d line, for "the discoveries" read: the discoverers.

On page 16, 28th line, for "Peterzen" read: Pieterzen.

On page 19, 4th line, for "catalogue" read: catalogues.

On page 55, 30th line, for "home and asylum of those who had deprived him of liberty and life" read: asylum of those who had deprived his people of liberty and life.

On page 63, 42d line, for "1852" read: 1822.

On page 107, 11th line, for "seventeen" read: seven.

PREFACE.

Every work should be justified by its usefulness and recommended by the manner of its performance.

Criticism of literary style is averted from this little book, since nice elaboration of details, and smooth, consistent unity of parts, with a high degree of literary finish, are impossible in a volume made diverse by the requirements of its purpose and desultory by needful brevity. I have been disinterested in that of which I have written, and left entirely free to follow my own taste and judgment in regard to matter and manner, being bound in agreement with those concerned only that I should serve their purpose by "truthful representations" alone.

Thus directed and encouraged in pursuing the course congenial to my feelings and conscience, I have tried to present only the facts of science and the truth of history, knowing them to be stranger than fiction, and in simple statement more wonderful and interesting than the most remarkable works of imagination.

Although an observer of the things I have described so far as they exist in the present, it would be absurd to put forward any claim to original discovery. I have gathered from many sources, but think a display of authorities would be out of place; yet, it is true, I have been more inquisitive than the result may indicate. Errors are possible, even when care is taken to be accurate, and mistakes are not at all inconsistent with an honest purpose; still, if misrepresentations exist in this work they are unknown, and as the motive has been conscientious, and the effort earnest, I believe the consideration due reliability is deserved by all herein published.

But whatever discrepancies may mar the printed pages, there is no occasion to criticise the illustrations for misrepresentation. They are mostly drawn from photographic views, taken on the spot, with microscopic fidelity, by artistic operators, and have been faithfully reproduced by the draughtsman and engraver. They may, therefore, be looked upon as giving a correct idea of the physical features of the beautiful locality in which they were taken, and the varied structures which utilize and decorate the neighborhood.

Whoever has loitered along the shore of the summer sea, seeking rest and recreation therefrom, has, when feeling his soul stirred by the grandeur and loveliness of the scene, longed for some magic art which could fix forever the transient glories of evanescent beauty in his mind, making his memory thus the picture-gallery of nature.

This may not be, but somewhat has been done to recall the features of the seascape where, in the bygone summer, so many earnest, Christian souls "took sweet counsel together," amid the healing breezes and peaceful surroundings of the consecrated Sea Grove.

Neither the artist's pencil nor the photographer's skill can reproduce all that presented itself before the delighted vision. No art can imitate the tenderness of the dawn across the sea, or do justice to the resplendence with which the sun sank among the western waves on quiet Sabbath evenings; but all this may be suggested to the sense, and with many memory will fill the picture with colors true to nature, and even recall the friends who shared their summer vacation.

Again, as they look upon the pictures of our unpretending book, they will hear in memory the voice of exhortation and the music of praise, mingling with the undertone of the unceasing surges. Again they will enter the broad pavilion, and, pausing but to offer a word of prayer for all who share not in their religious blessings, bow the soul in devotion to partake of the union communion service with their numerous friends of many churches.

To awaken such reminiscences in those who know Sea Grove and its associations by residence there, and to increase their interest and pleasure in the place by bringing before them many facts pertaining to their favorite resort, is the purpose of this book; besides, it is requisite that all who need the sea-side privileges of rest and cheerful recreation should be informed where they can secure them at their convenience, reasonably, without annoying contact with demoralizing dissipations, as distasteful to the thoughtful as they are wearisome and hurtful to the invalid, and physically and spiritually unprofitable to all.

Trusting that these ends may be fully served to the common benefit, and that something of instruction and refined gratification may be incidental thereto, the author with pleasure presents his work to an enlightened public.

ORIGINAL LIGHTHOUSE AT CAFE MAY. REMOVED IN 1847.

SCHEYICHBI AND THE STRAND.

HISTORY evinces the exceeding potency of religious ideas, as a cause of material progress; as the phenomena of Nature manifest the power of the Infinite Spirit.

Curiosity, avarice, and ambition induce exploration and discovery; stimulate enterprise; found and foster states; but fanaticism, faith, and spiritual convictions are the world's pioneers; these move more profoundly the passions of mankind, quicken higher and intenser energies, and develop more sublime results.

Fanaticism, the *fungi* of religious growth, provokes the bigot to draw the sword of exterminating conquest, changing the character and boundaries of nations; the mad zealot lights the fires of persecution, expatriating the flower of a country's population, who carry religion and the arts into their place of banishment. Devotion inspires the propaganda, and missionaries penetrate the antipodean wilderness, domicile among barbarians, and plant civilization to flourish above their martyr graves. Faith feeds the courage of the believer, and impels to self-consecration; fired by religious enthusiasm, bound by stern conviction, and led by the "inward light," the dissenting Huguenot, the Covenanter, the Puritan, and the Quaker dare the ocean, the desert, and the savage, in search of a home of righteousness, for freedom and for peace. Hope stimulates them, a religious purpose sustains them; they confront every peril, endure every trial, survive all suffering, outlive every hinderance, and triumph at last over every difficulty in the adorable name of God!

§ Prophesied in the rhapsodies and inspirations of the seers of all ages; mysteriously reported in the literature of Asia in the early dawn of the Christian era; celebrated obscurely in the historic runes of the heroic Scandinavian sea-kings a thousand years ago, and claimed by Icelandic and Danish historians as the familiar haunt of their forefathers for many centuries,—the Western Hemisphere long nourished on its soil nations who imitated the architecture of Egypt, perpetuated the religious rites of Tyre, and may have shared in the commerce of the Orient. On the shores of the Western World, it has been claimed, was mined the gold of Ophir for the temple of Solomon; while the broad plains of its continents received, it is said, the lost and wandering ten tribes of Israel.

Reflecting dubiously the life of unknown ages, from the sculptured sides and hieroglyphic ornaments of its antique and symbolic monuments, America inspires the imagination, but compels the mind to drift unsatisfied over its vast and significant ruins, back into the twilight of tradition and the night of pre-historic oblivion. The plains of America are marked by the work of a race without a record; its great valleys covered with traces of a numerous and active population, and yet they have no chronicle. The American forests tower above the ruins of large cities whose civilization is evident from their architecture,—still the hosts of citizens have passed away: their origin, their history, and their fate conjecture alone can intimate.

But, if the past of America is perplexing to the antiquarian, dubious in historic twilight, or hid in the darkness of time and barbarism, its modern life is clearly defined and of thrilling interest. Here are no monuments of an enduring civilization, linking the present, generation by generation, to the remote past; no vast collections of splendid volumes, the record of a people's ancient glory; no empire, one in faith and one in government for a thousand years,—all is new, primitive, incomplete; but there are young states in America proud as Rome, more free than Athens; there are a hundred great, luxurious, and growing cities; there are public works that open up the long sought passage to India, and millions of happy homes, of the best provided, most intelligent, free, and independent people.

It is less than four centuries since the voyages of Columbus; the history is brief, but the advance has been rapid, the development immense. Each American generation has done the work of a hundred years, and each century has become an era in civilization, an epoch in history. To compile and elaborate the record of such an advance, and educe the principles of progress from the facts of social and political evolution, is the congenial and proper work of philosophic scholars, and acute and comprehensive minds have employed themselves therein with usefulness and honor.

It is not the purpose of the writer to ape the great historiographers, but he may modestly hope to add a reliable note to the materials of history, suggest some practical inference, or inspire an appropriate reflection, just as the wandering but observant Indian, though unskilled to build the monument of a nation, still faithfully places a votive pebble upon the growing mound which tells of the greatness of his tribe.

However little the present publication may add to the vast sum of historic knowledge, it at least indicates the causes which have fostered American liberty, and manifests the nature and temper of a free people as the energetic cause of moral improvements and unexampled material progress; this appears in the history herein given of the settlement of the valley of the Delaware, especially in New Jersey, and conclusively in the interesting and detailed account of the developments of Sea Grove,—that beautiful and prosperous town having been instituted entirely in keeping with the spirit of the representative men composing the Association which bears its name.

§ The discovery of America was prehistoric; its unrecorded monuments, ruins, and sculptured rocks were antiquated when, in 1492, Columbus voyaged to the West Indies, and various nations and races had already left the traces of their visits and occupancy at a number of widely separated localities upon the two Western Continents. The modern history of America begins with the voyages of the inspired navigator of Genoa. The rediscovery of the Western Hemisphere commanded the attention of the civilized world; aroused the emulation of nations, and the ambition of kings; it inflamed the spirit of the adventurous and enterprising; kindled the imagination of the enthusiastic; awakened the hopes of the people; encouraged the aspirations of liberal statesmen, and actualized the dreams of the philanthropist.

India was the prize Europe coveted four hundred years ago. Columbus sailed for Cathay, and supposed he landed on its eastern shore,—"the beginning and the end of India." His voyages for a short route to India discovered America; the search for a northwest passage explored the shores of the "New World."

In the time of Columbus it was the uncertain international law of Christendom, that Christian nations became entitled to any land or country [illegible] citizens discovered, took possession of and occupied, unless it was already the territory of other Christians. This presumptuous claim of the exclusive right of a sect, as such, to the secular ownership of the whole world, was a political device, and, though endorsed by popes and approved by bishops, was at once absurd, impudent, and irreligious; but the heresy had a natural origin, and, becoming a dogma and an apology, developed an awful historic sequence.

Numerous as the voyages of discovery to America were, and important as trade became, for more than a hundred and fifty years after Columbus, gross ignorance of the Western Hemisphere characterized the action of even the courts and kings of Europe. Under the name of the "West Indies," two vast and rich continents were long regarded as but troublesome islands in the way of voyages to India, and frequent and conflicting royal grants afterwards assumed to convey, in an impossible manner, possession of the territories of America from ocean to ocean, the grantors having the untroubled conceit that the average width of the continent was no more than about three hundred miles.

Under the pretext supplied by the voyages of Columbus, Alexander VI., "the worst of the popes," assuming to be the temporal as well as spiritual head of Christendom, pretended to invest Spain with regal possession in perpetuity of all heathen lands found, or to be discovered, to the west of a meridian three hundred and seventy leagues westward of the Azores. In insolent and fanatical assertion of her declared rights, which, thus derived, became a matter of religious faith, Spain undertook to monopolize the trade of the West Indies and control the navigation of the high seas. Hence, Portugal colonized and traded only in part of Brazil, her minute allotment of all the vast "Indies;" and so, in defense of the faith enshrined in her Papal monopoly, the fleets of Spain pirated all vessels they overhauled sailing the Atlantic to her pretended exclusive possessions. At the same time Spanish kings made war upon Protestant maritime nations in a way that left enterprising Holland no chance for existence but in her defeat, and compelled England to sail to commercial and naval supremacy over the sunken hulks of the "Invincible Armada."

§ Although Balthazar Moucheron, of Holland, and his associates, patrons of discovery, moved by the terrible sufferings and failures of their explorers, about the year 1600 abandoned as hopeless the quest for a northern route to India, the immense importance of such a passage was obvious, and the Danes and English continued the resolute search. The directors of the prosperous and powerful Dutch East India Company, then in full operation, shared the notions of their cotemporaries, and, overruling the experienced Moucheron and his Zeeland partisans, the Amsterdam members of the Directory, jealous of Denmark and England, decided the Company to seek for itself a safer and more convenient way to their remote places of traffic. The stockholders of the East India Company had received in one year a dividend of seventy-five per cent. on their investment; they could well afford a venture which promised even greater facilities to their business. By orders from the Directory at Amsterdam, a very fast sailing vessel named "De Halve Maan," or Half Moon, of forty lasts or eighty tons, a "vlie-boat," having two masts, such as were constructed especially for difficult navigation in sounds and rivers, was fitted for an arctic voyage. For a schipper, or commander, Henry Hudson, an Englishman, who had already made two such adventures, was engaged. The under schipper, or mate, was a Dutchman, and the vlie-boat was manned by twenty men, English and Dutch. Robert Juet sailed with Hudson as his clerk, and became the historian of the voyage. The De Halve Maan was ordered to look for a passage by the northeast or northwest to China, the Directors trusting Hudson to find some way past Nova Zembla, or some strait or channel between the islands of the West Indies, by which their fleets of Dutch East Indiamen, fearless of Spanish interference, could bear directly to India and all the Orient the products of Europe in profitable exchange for the pearls of the Asiatic Archipelago, the diamonds of Golconda, the lawns of the Deccan, and the spices of Cathay.

Accompanied by his only son, Hudson, hailing from Amsterdam, set sail the 4th of April, 1609, for the northeast of Norway. He left the Texel on the 6th of April, and doubled the cape of Norway on the 5th of May. Finding his way toward Nova Zembla obstructed by vast icebergs, and his ship crowded out of her course by great fields of moving ice, Hudson ran the Half Moon to the west and south. Passing through a great fleet of French fishermen off Newfoundland, and touching at several points on the coast of New England, he arrived off the Chesapeake in the middle of August. Hudson's old friend, Captain John Smith, had given him a map of Virginia, on which, somewhere to the north of the Chesapeake, a strait was laid down, by which Smith was confident the Pacific Ocean could be reached. Knowing himself to be in the neighborhood of the settlement of his countrymen and friends at Jamestown, Hudson put his ship about, August 18th, and kept along the coast to the north again. The Half Moon entered Delaware Bay August 28th, which Hudson slightly explored and sounded, making observations of its shores, but without landing. Finding he could not sail his vlie-boat from Sea Grove to San Francisco, and hence that the Delaware was not the passage to Cathay, Hudson coasted to the north along the Jersey shore, and on the 3d of September anchored inside of Sandheuken, or Sandy Hook, where he remained a week, and was frequently visited by the Indians. From this anchorage the Half Moon sailed into the bay of New York, still being visited by the Indians, whom Hudson and his crew taught, as their first lesson in civilization,—how to get drunk.

Hudson examined the Hudson River for twenty-two days, his boats going up twenty-five or thirty miles above Albany, and then, having made sure that neither Hell Gate nor the Hudson were a water-way to Hindustan, he, on the 4th of October, put out to sea, and, in consequence of the dissensions of his crew, finally decided to set sail for Holland.

The Half Moon with her motley and mutinous company, of whom Hudson became afraid, put into Dartmouth, in England, where, the Dutch assert, she was detained and Hudson kept through the jealousy of James I. Hudson, however, sent a brilliant report of his voyage to his employers in Holland, in which he speaks of the country he visited as "most beautiful," "het scoonste land dat men met voeten betreden kon," etc. Whoever has voyaged up the "Great River of the Mountains," above New York, by the Catskills, or yachted in August off Sea Grove and up Delaware Bay, where the vlie-boat De Halve Maan cruised in that month long ago, will certainly agree with him.

During his fourth voyage of discovery, made from England in 1610, Hudson with his only son and eight men, four of them being sick, was driven by mutineers from his ship, the Discovery, into an unprovisioned boat and cast loose among the ice, mid-seas in Hudson's Bay. There the brave and persistent navigator must have cruelly and miserably perished.

Could he but voyage once more out of the cold and ice-bound Arctic seas, how overwhelming would be his astonishment! At the extreme point of Cape May he saw, with admiration, long ago, the green woods crowd down to the sandy strand, and from the primeval forest the wondering Kechemeches stare out, thinking his ship the canoe of their Manitou. There he would now look in amazement upon the broad avenues and handsome cottages of Sea Grove; he would see hotels and pavilions in the place of savage wigwams, and hear the Sabbath bell, the organ, and the Christian hymn, instead of "the gaunt wolf's long-drawn howl" along the shore, or the war-whoop of the exultant savage.

"The bay of the south river was the first place of which the men of the Half Moon took possession, before any Christian had been there," says Vander Donk, the historian; and the claim of the Dutch to the adjoining territories by right of discovery was based upon the assumed accuracy of the statement. Hudson may have been the first to formally take possession of the Zuydt Baai, as the Hollanders called the bay of Delaware, but Cabot, Cortereal, Verazzani, Captain John Smith, and others, had at various times carefully observed the shores and harbors of "Virginia," and cruised along the coast to the north; besides, it is historical that very early, scores of years before the voyages of Hudson, "there was hardly a convenient harbor on the whole Atlantic frontier of the United States which was not entered by slavers." It seems that Hudson, following, perhaps unconsciously, in the wake of others, merely took possession of the unrecorded discoveries of some unknown navigator.

§ In answer to the petitions of a number of merchants, a general edict was issued by the States General of Holland, March 27th, 1614, for the encouragement of discovery and the protection of aboriginal trade. It was enacted by the High and Mighty States General that the discoverers of "any new courses, havens, countries, or places" should have "the exclusive privilege of resorting to and frequenting the same for four voyages," and all intruders were to be punished by confiscation and fines. A number of merchants, chiefly of Amsterdam, thereupon formed a partnership to make discoveries and carry on trade to new countries, and five vessels were fitted out to follow in the track of Hudson to Manhattan. One of these, named the Fortune, was from Hoorn, a port in North Holland, and commanded by Cornelis Jacobsen Mey; another ship, also called the Fortune, was in charge of Commander Hendrick Christiaensen; a third, named the Tiger, was sailed by Captain Adriaen Block. Arriving at the mouth of the Hudson, Block's vessel was accidentally destroyed by fire. To retrieve this misfortune, he erected a few huts at Castle Garden, and began to construct a yacht of about sixteen tons burthen, of the fine timber he found there, the Indians kindly feeding him and his men, all the winter of 1613. May, in the mean time, cruised to the eastward, coasted along the southern shore of Long Island, and continued his trip to Martha's Vineyard, then called "Capacke" by the natives. Upon the completion of his new craft, the Onrust, or Restless, Block sailed through the East River and Hell Gate, where he led the way as a pilot, and through Long Island Sound, observing the coasts, harbors, islands, rivers, and waters, as far as Cape Cod, the promontory to which Hudson, in the summer of 1609, had given the name of "New Holland." Block ascertained that Long Island was sea-girt, and visited many other remarkable places along the New England coast. The records of the voyages of the consort ships, the Fortune, the Little Fox, and Nightingale, in 1613 and 1614, are imperfect and unreliable.

The name of Block Island perpetuates the memory of its persistent and intrepid discoverer, the first man to run a keel through Hell Gate, and the first "Long Island Sound Pilot." The shores which Block surveyed, and which Holland first colonized, have been for two centuries or more, as now, "the land of steady habits," the home of industry, prosperity, intelligence, and freedom,—a "New Holland," indeed, a "New England" as well. They are glorious by day with many a fair town and city, and sparkle at night with scores of shining beacons, while over the seas the Dutchman slowly navigated speeds in ceaseless succession a numerous fleet of "floating palaces," the best, the safest, and most magnificent steamboats in the world.

The "Restless," built at Manhattan, in 1614, was thirty-eight feet in the keel, forty-four and one-half feet from stem to stern, and eleven and one-half feet wide. She was remarkable as the first vessel built in the harbor of New York, but was not, as has been written, "the first decked vessel built in the old United States," the "Virginia," of "Sagadahoc," of thirty tons, a "pretty pinnace," having been built by "one Digby, of London," at St. George's,—Sir George Popham's settlement, —at the mouth of the Penobscot River, in the winter of 1607. Still, the Restless was a notable craft, for she sailed in the van of a countless fleet, which for two hundred and fifty years has stood out from the northern coast of the United States to astonish the navigators of Europe by the excellence of American ships, and furnish models for the improvement of the naval architecture of the world. Of all the many fine ships which have done honor to American shipwrights, a creditable share have been launched in the waters of the Delaware. Since the "iron age" of shipbuilding, the craftsmen of its shores have made their names honorably known from London to the "city of Pekin," and now compete with England and Scotland for supremacy in trade, confident of surpassing the industries of the Clyde on the banks of the Delaware.

The Restless explored her way to "Pye Bay," now Nahant Bay,

Massachusetts; there she fell in with Christiaensen's ship, the Fortune, also on a cruise. Leaving the Restless in command of Cornelis Hendricksen, to be used in exploring on the coast and in the rivers, Block returned to North Holland and made his report to his employers. From his sketches and descriptions an elaborate "Figurative Map" was made, and laid before the States General, with a request for a charter for those who had procured the discovery of the lands delineated upon it, without delay. A special grant, dated October 11th, 1614, was made to the Amsterdam partnership; they were conceded the monopoly of trade from forty to forty-five degrees north latitude on the coasts of America. The partnership took the title of "The United New Netherland Company." The territory assigned them was called New Netherland. At the same time, at Manhaddoes, or Manhattan, their principal fort was named "New Amsterdam."

The first vessel built at Manhattan was the first to cruise the Delaware. Hudson, in 1609, was too fearful of getting aground to attempt explorations in Zuydt Baai, though less timid in the Noordt Riviere. Argall, on his return from his mysterious cruise in 1610, remained but a day at anchor in the Delaware, leaving the same evening for the Chesapeake, but, in 1616, circumstances led to an exploration of the Poutaxit. It happened that three fur traders, agents of the New Netherland Company, having left Fort Nassau (near Albany), and made their way along Indian trails to the mouth of the Schuylkill, were there kept prisoners; news of this reaching Manhattan, the Restless was sent from the Mauritius River, under command of Cornelis Hendricksen, to ransom the adventurous captives. Block had constructed the Onrust for shallow waters and inland navigation; so Hendricksen, on his arrival at Zuydt Baai, coasted fearlessly along the western shore, making careful observations, bartering with the natives for seal-skins and sables, and being delighted with the scenery, climate, and vegetable productions of the valley, until he arrived at Coaquannock, "the place of tall pines," now central Philadelphia; there he found and ransomed his countrymen for "kettles, beads, and other merchandise."

The people at Manhattan now called the Delaware River New, South, or "Zuydt" River, and the southern Cape of Zuydt Baai, now called Henlopen, was soon known as Cape Cornelis, after Cornelis Hendricksen. A point some miles south of Cape Cornelis was named Hinlopen, in honor of Thymen Jacobsen Hinlopen, of Amsterdam, one of the "Northern Company," engaged in the whale fisheries and explorations, by which Block was employed on his return from America. Cape Hinlopen was also called Inloopen by the Dutch schippers, because it seemed to recede from sight when approached from the sea. The names of these capes have been transferred, and the name of Henlopen is now borne by the point at first named Cornelis.

§ In 1620, Cornelius Jacobsen May, who in 1614 commanded the Fortune of Hoorn in the explorations along the coast east of Manhattan, came again to New Netherland in a new vessel called the "Blyde Boodschap," or Glad Tidings. This voyage was intended for the exploration of territories to the west of and below Manhattan, and those south of the fortieth degree to "Virginia," and was made to include Zuydt Baai and the Chesapeake, which the Blyde Boodschap ascended, and went up the James River to Jamestown. May carefully examined the bay and river of the Delaware, where Hendricksen had preceded him four years before, and then returning to Holland early in the summer of 1620, announced the discovery of "certain new populous and fruitful lands" along the Zuydt Riviere. The Poutaxit, Zuydt, or Delaware Bay, as the Indians, Dutch, and English had named it, was after this called "Nieuw Port Mey," and the name of "Cape Mey" was given to the southern point of New Jersey, then as now "the best bathing place in the world."

May, as Hendricksen had done, indeed as every one does who visits Cape May in summer, found the climate charming. It was the highest compliment they could imagine, when the Dutch explorers, a home-loving though voyaging people, declared the climate of the Delaware was "like to that of Holland;" as good as home. As it happened, both the nomen and cognomen of Cornelius Jacobsen May were applied to capes at the mouth of the Delaware, but the name of Cornelius, given in honor of Hendricksen, has been thrust aside and made insignificant, while the fame of CAPE MAY has become world-wide, and summer by summer its increasing attractions add to its popularity, as time multiplies its appreciative visitors.

§ The principles of the Lutheran Reformation gave permanence and character to the colonization of the United States; the hand of persecution pointed the way to New Netherland, and the valleys of the Hudson and the Delaware became an asylum from ecclesiastical despotism even while the Puritans of New England, jealous of their own freedom, denied liberty to others. When, in 1623, the great Dutch West India Company, complete in organization, sought to people its territories, the victims of persecution offered themselves as its first and most desirable emigrants.

When the Hollanders, after their revolt against Spain and the Inquisition, in 1565, formed the Union of Utrecht, the Belgic provinces of Hainault, Namur, Luxemburg, Limburg, and Liege, having mostly Roman Catholic citizens, did not join the Dutch Confederation; still, many of the Belgic people were Protestants, and as such were victims of persecution under Philip II. of Spain. Speaking the old French language, these people were termed Gallois; they fled by thousands to Holland, where their skill as well as their faith secured them protection and a welcome. In low Dutch the name of the refugees became

"Waalsche," which the English rendered Walloons. The farmers among the Walloons found poor encouragement in Holland, and in 1622 a number of them offered to emigrate to Virginia if assured municipal freedom. Some delay followed their application to the British Minister at the Hague, and meantime those willing to form a settlement were, at the suggestion of the Provincial States of Holland, engaged as colonists by the Dutch West India Company, and employed in Holland until such time as the perfect organization of that corporation would enable its Directors to send the Walloons to New Netherland.

Having by virtue of their charter taken possession of their domain in 1622, the Dutch West India Company secured the assent of the States General to their articles of internal government the 21st of June, 1623. The same month three trading ships were dispatched to Manhattan "to maintain the course of traffic," and a special effort was made to colonize "Nova Belgia." The "New Netherland," a ship of two hundred and sixty tons, was fitted up, and on board her were embarked a company of thirty families, mostly the Walloons who had offered to settle in Virginia. The superintendence of the ship and colony was entrusted to Cornelius Jacobsen May, who was appointed to remain in New Netherland as First Director; his second in command on the ship being Schipper Adriaen Joris, of Theinpont. The expedition left the Texel early in March, and, following the southern route by the Canary Islands and Guiana, came in safety to Manhattan, the beginning of May. At the mouth of the North River the emigrants repulsed a party of Frenchmen, who were about to erect the arms of France; the French ship, however, renewed her attempt at Zuydt Baai, but was driven off from there by the Dutch settlers or traders. At an early date the Dutch established a lookout at Cape May, and from the time Cornelius Hendricksen in the Onrust explored the Delaware, they were generally well informed of whatever took place thereabouts, and frequently warned off whoever entered.

At Manhattan, Director May left several families, and a number of sailors and men from the New Netherland, for the settlement of South River and the shore of the sound eastward. The ship then proceeded with difficulty up the North River, and landed her company just above Castle Island, on the western bank of the Hudson, at Albany. There "a fort with four angles, named Orange," which had been plotted the year before, was soon completed; the industrious Walloons "put the spade in the earth," and when the next yacht sailed for Holland, their corn "was nearly as high as a man, so that they were getting along bravely." Brave hearts, heroic souls, the verdant corn you tilled struck no root so deep in the soil of the New World as the faith for which you were exiles, no harvest spread so rich a growth as the principles of freedom and toleration you planted here! Down the Hudson every year floats the wealth of granaries, richer than Egypt, but the spirit of Religious Liberty and Civil Independence, entrenched in the hearts of millions, bids defiance to intriguing priests and threatening tyrants as it breathes the benediction of "Peace on earth and good will to men" over the vast expanse of a mighty continent.

To prevent attempts to occupy Zuydt Baai, the fort projected in 1622 was, by order of Director May, speedily completed. It was built five miles below Philadelphia, on the Jersey side of the river, of great logs, and named Fort Nassau, the first post of that name, on the island near Albany, having been destroyed by flood and ice. There were four weddings on board the New Netherland during her two-months' voyage from Holland over the sunny Southern seas. Director May, who was a kindly man, had been directed to govern his people "as a father, not as an executioner;" and it was with a touch of romance, as well as paternal care, that he selected these eight newly-married Walloons, and sent them, about the first of June, in a yacht, with as many sailors, to abide at Fort Nassau. They were far from home, from friends, even from civilization, a mere handful in the wilderness among savages, but they were enough; each for the other of every pair, and all for each of the quadruple family. It was a fitting and poetic thing that the valley which was to welcome the men of peace, and grow in peace to be the home of freedom, should owe its first historic settlement to young and joyous brides, with their free and hopeful partners. It was in harmony also that they should come in the freshness of summer, when the very air was balm, when every leaf told of life and vigor, when every forest aisle was sweet with woodland fragrance and echoing with bird songs, every note swelling the all-pervading melody, one perfect chorus, whose glad refrain was evermore of love, and still of universal, all-embracing love.

Eighteen of the Walloon families settled at Albany, others went for a time to the House of Good Hope, at Hartford, Connecticut; others made themselves homes, in comfort and happiness, on Long Island. There, in June, 1625, Sarah Rapelje, the first white child of New Netherland, was born; and thereabouts, in usefulness and honor, the descendants of the Calvanist *Gallois* still reside.

§ Of Cornelius Jacobsen May, who was formally installed during the summer of 1623 as the first Director-General of New Netherland, there is but little more to be said, but that little is entirely to his credit. "'Tis better to govern by love and friendship than by force," wrote his superiors in Holland; and May acted in the spirit of his instructions, to "the great contentment of the people." Among the Indians at Fort Nassau May's little colony of brides and grooms were unharmed, while at both Manhattan and Fort Orange the Indians "were all as quiet as lambs, and came and traded with all the freedom imaginable." It required other men than May, and other means than "love and friendship," to arouse the savage in the red man of America.

Mentioned as a man of experience at the time of his appointment, Director May had many unrecorded adventures. During one of his earliest voyages to America he found the colonists at Manhattan suffering for stores and clothing. From his own ship he supplied their necessities, and the grateful Manhattanese celebrated the timely relief by giving the name of Port May to their harbor.

The voyage of Director May to the Delaware, in 1620, was commemorated by the name of New Port May, applied to the bay of the Delaware, and by that of Cape May, ever since retained by the southern point of New Jersey.

Thus circumstances supposed to indicate the vanity of May in affixing his name to various localities are explained either as a just tribute to the deeds of another exploring "Cornelius," or the grateful and graceful act of his people.

Cape May is one of the very few points about the Delaware which retain the names first given them by white men; but of the thousands who visit it annually, very many are not aware of the source from which that name was derived. Some, careless of history, infer from their pleasant experience of its balmy atmosphere that Cape May derived its appellation from the May-like breezes which make its summers "balmy as the breath of spring." But "the Cape," especially since the improvement of Sea Grove, has too many charming attractions to need misrepresentation to make it popular. By nature and improvement Cape May is superior as a seaside resort, but its name is significant only as a memento of the old-time voyages of the Hollanders, and of their regard for the character and exploits of their popular Superintendent.

Though the name of Cornelius Jacobsen May disappears from this history, the admirers of Cape May have reason to be proud of the name it bears, since it recalls only deeds of courage and goodness, such as confer an honest fame in the history of time, and crown with happiness the pure in heart amid the glories of eternity.

§ But while perfect peace and fair prosperity marked the history of their colonies, the Directors of the Dutch West India Company were disturbed by the enterprise of a person destined to play an important part in the events of New Netherland. A mariner of Hoorn, North Holland, by the name of David Pietersen De Vries, who had several times voyaged to Newfoundland, procured a commission from the King of France, and, dividing his venture with some Rochelle merchants, he bought a small vessel for a voyage to Canada, for fish and peltries. Determined to prevent all ships but their own sailing to North America from Holland, the Directors seized the vessel of De Vries as it lay in the harbor of Hoorn ready to sail, and detained it until an admonitory mandate of the States General ordered its release. De Vries re-

SEA GROVE

VAN INGEN=SNYDER

RUSTIC GATEWAY TO SEA GROVE. LOOKING SOUTH.

ceived his vessel after much delay; although his voyage was broken up, his claim for damages was evaded, and, suffering from corporate injustice, the enterprising navigator was compelled to bide his time and await another chance of fortune.

Director William Verhulst presided over New Netherland in 1625. He visited the Delaware and extended his voyage far up to the falls at Trenton; there on an island in the bend of the river another trading post was established, and for a time occupied by several families of Walloons. Verhulst returned to Holland in 1626, and Peter Minuit became Director-General of New Netherland. "To superadd a higher title" than that supposed to be derived from discovery and occupation, Minuit purchased Manhattan from the Indians. The island contained about twenty-two thousand acres, and was bought of the natives "for the value of sixty guilders,"—about twenty-four dollars.

Having bought Manhattan, the Dutch began a fort, "to be faced with cut stone," for its defense; for the misbehavior of some of the colonists had given reason to fear just hostility. About the same time the posts on South River were much reduced, and in 1628 left untenanted, in order to strengthen Manhattan. Still there were in all probability settlers left on the Delaware, not perhaps the servants of the Company, but "vrye persoonen," who had reason to trust their Indian neighbors, and led a roving, adventurous life among them; but of these adventurers history, made up of corporation documents, has nothing to relate. Before the completion of the fort at Manhattan, it was called Fort Amsterdam, and made the seat of government. There has been a great amount and variety of government on the island of New York since that date, and not a little misgovernment; but with it all an undeniable increase of trade, and a most notable advance in the price of real estate.

§ The United Provinces of the Batavian Republic elaborated the idea of federal union, but their institutions failed to develop personal liberty; the peasantry of Holland had therefore too little self-reliance to emigrate, and a plan was evolved to encourage colonization, called the Charter of Privileges and Exemptions. By the provisions of this new charter of 1629, whoever of the stockholders of the Dutch West India Company established a colony of fifty persons within four years in New Netherland, became a "Patroon" or "Lord of the Manor." The Patroon had jurisdiction over the settlement he founded, and, by peaceful purchase from the natives, might hold and own the lands on the sea-shore or river-bank for sixteen miles, and as far inland as "the situation of the occupiers would admit;" or the land each side of a river could be held half as far, with a *pro rata* increase for more colonists in each case.

While the Charter of Freedoms and Exemptions was under consideration, several directors of the Dutch West India Company, tempted by the concessions it made, undertook to forestall its provisions, and embezzle for themselves in advance the richest territories of the corporation. These crafty schemers sent three ships to America with agents to locate manors, and buy the land of the Indians. One of these ships entered the Delaware in May, and on the 1st of June, 1629, a few days before the adoption of the charter in Holland, "two persons," who came on the ship, bought for directors Samuel Godyn and Samuel Blommaert, from the natives, a tract of land two miles wide, which extended from Cape Henlopen thirty-two miles up the bay to the mouth of the river.

At the first meeting of the Amsterdam Chamber after the adoption of the Charter of Freedoms and Exemptions, Director Samuel Godyn gave notice that he as Patroon occupied the bay of the South River, having notified Minuit to register his possession of the same at New Amsterdam. To remove the dissatisfaction which was manifest in the chamber at the course taken by them, and to secure capital, the Patroons admitted Killiaen Van Rensselaer, Johannes de Laet, the historian, Mathias Van Ceulen, Hendrick Hamel, Johan Van Haringhoeck, and Nicholas Van Sittorigh, as partners in their enterprise. In order to secure his services as superintendent, Pieterzen De Vries was made an equal partner in the concern. The ship "Walvis," or Whale, carrying eighteen guns, and a yacht, were fitted out at once for an expedition to the Zuydt Baai. The two vessels were loaded with colonists, stock, animals, seeds, tools, and the requisites of an agricultural colony. At the suggestion of Godyn, implements were also taken for the capture of the whales, seals, and sturgeons, then abundant in the Delaware.

Amply supplied, the expedition left the Texel December 12th, 1630, under command of Pieter Heyes, of Edam, North Holland, Peterzen De Vries remaining in Amsterdam. Through carelessness on board the Walvis, the yacht was captured by Dunkirk privateers, but the ship kept on, and, passing by Tortugas, where a part of her colonists were bound on French account, but which was found in Spanish hands, she completed her trip. In April, the Whale arrived safely at Zuydt Baai. Finding a safe landing and convenient harbor, with islands, good oysters, and very fertile land, the colony was landed up the stream on the banks of a "kill" (creek, or small river), near the present Lewes, Del. This stream, which was called after the city of Hoorn, Hoornkill, Hoorkill, etc., afterwards corrupted to Whoorkill, or Whorekill, was also called the river of Swans, and was reported to be two leagues from "Cape Kornelis," now Cape Henlopen, the site of the splendid light that, with its equal and neighbor at Sea Grove, illuminates the wide entrance to the Delaware. In the vale where the Dutch colonists landed there were many swans, and hence they gave their settlement the name of Swaanendael (Swandale).

Gillis Hossett, a former agent of Van Rensselaer's in the purchase of lands from the North River Indians, was placed in command of the station; a large brick house was built of Holland brick, and enclosed with palisades; this building served at once as a residence for all the colony, a storehouse, and a fort. As soon as the settlement was well begun, Commissary Hossett and Schipper Heyes visited the Jersey shore, and, as agents of Godyn and Blommaert, bought of ten Indian chiefs, on May 5th, 1630, a tract of land twelve miles along the shore of the bay, from Cape May Point to the north, and twelve miles inland above, and including Cape May. The lands on the northern and eastern shores of Delaware Bay were in possession of the great and influential but peaceable tribe, called Lenni Lenape (the original people). From them must have been obtained the original title to Cape May; and the Nanticokes, who occupied what is now Delaware, must have been the grantors who, on July 15th, 1630, ratified by treaty the sale of the western shore of the bay, made to Godyn and Blommaert's agents the year before.

§ Such is the record of the first transaction in real estate at Cape May; the advance in value on the smallest building lot in Sea Grove, for the current year, is represented by a sum of money greater than was needed to buy the lands of all the lower Delaware; yet both parties were well pleased with the speculation. The Indians, who knew but little more of the full purport and effect of a deed of land than the deer of the primeval woods, were delighted with the "presents" they received, and charmed by the civil and novel manners of their liberal customers. The patroons needed but to examine their purchase to become satisfied they had come into possession of a land of promise. Zuydt Baai was now called Godyn's Baai, by which name it was afterwards well known to the Dutch. After spending a few weeks at Swaanendael, Heyes, with Hossett in company, visited New Amsterdam, and there, on the 3d of June, 1631, had the purchase they had effected formally recorded and attested by Director-General Minuit and his council. The deeds of the lands purchased on the Delaware for Godyn and Blommaert were deposited at Fort Amsterdam, and conveyed to Holland, but are now in the archives of the State of New York, at Albany.

Whaling was undertaken by Heyes in Godyn's Baai, but the experiment was a failure, and, in September, 1631, the Walvis sailed for Holland. Gillis Hossett remained at Swaanendael to superintend that colony, and, by more thorough explorations of the new manors and their resources, prepare the way for future settlements.

Pioneer explorations must have been magnificent in those days. As Hossett sailed over the waters of the Delaware he saw a roadstead and harbor, where all the commerce of Europe could ride secure; the low shores on either side reminded him and his companions of Holland, as they offered every facility for the construction of canals, in broad marshes, which could easily be redeemed from the sea, and turned into fertile fields. But, unlike Holland, Cape May had dense forests of varied timber near the shores, for the countless hulls of navies, such as the world had never seen, and beyond, yet near, interminable swamps where the giant cedars towered,—an arsenal of imperishable planks and spars to equip every craft, though each of them were more huge than ever sailed the Texel, or startled the dreams of shipwrights beside the Zuyder Zee. The waters swarmed with fish: the whale, the porpoise, the sturgeon, and the cod abounded; besides, there were black fish, blue fish, "green" fish, "silver" fish, and "variegated" fish; there were mackerel, gar-fish, drum, bass, perch, herrings, flounders, turbots, soles, eels, anchovies, mullets, porgies, smelts, and shiners, all affording "an ocean full" of excellent food; then there was also the flying-fish, and scores of other varieties more curious than eatable.

There is no historic evidence that Gillis Hossett or the mariner Peter Heyes tarried to catch all these kinds of fish; if not, it was their own fault; the fish were there, and one summer, just two and a quarter centuries later, Secretary Spencer F. Baird, of the Smithsonian Institute, caught them all except the whale, the sturgeon, the porpoise, and the cod-fish, in the waters near Cape May; moreover, any visitor of Sea Grove may have the same pleasure. The whale rarely visits the Delaware now; the porpoise still rolls lazily against the tide, but the sturgeon are comparatively few; yet if any transient dweller by the sea despises the capture of the smaller fry, and aspires to wage war upon veritable monsters of the deep, he can, by taking passage for deep water on the yacht of the Sea Grove Association, not only enjoy a trip over genuine ocean billows, but may, if favored by St. Peter, return with "a string" of *sharks*, and an appetite like that of the marine outlaw he captures.

At the time of Godyn's purchases, the marshes of Cape May were much more extensive, and the sounds and thoroughfares larger. The explorer found the inland waters of Cape May abounding with fine oysters, clams, crabs, and other shell-fish, as at present. The marshes around the sounds, and the savannas or slashes between the sandy beaches, were the haunts of countless water-fowl, some remarkable for their large size and notable appearance, while many of various kinds were estimable as game birds and known to the natives then as delicious delicacies, as well as to the sportsman and *bon vivant* of the present. In their proper season the Canada geese were immensely numerous, and their habitual resorts were also frequented by more than two dozen varieties of duck and plover, in flocks or pairs, by tens of thousands; among them was the world-renowned "canvas-back" (*Anas valisneria*). The meadows, marshes, and shores were overrun by snipe and loons, woodcock, rail, curlew, bitterns, herons, sand-pipers, and tern. Eagles, cormorants, hawks, gulls, and other fish-loving varieties of birds hovered over the waves and the quiet waters for prey, or, pirate-like, plundered others of the scaly prize. On the uplands the variety of birds was vastly greater—quite too numerous to mention outside of scientific catalogues The bald eagle, and ten or a dozen kind of hawks, half as many owls, and eight or ten kinds of fly-catchers exercised their capacity upon their varied and proper game; while the turkey-buzzard, with the help of several kinds of crows, was the common scavenger of the land.

Master Evelyn, William Penn, and others mention wild turkeys of the Delaware country which weighed from forty-five to fifty pounds. Grouse, partridges, pigeons, doves, and robins were abundant. Of birds of song there was no lack. There were fourteen kinds of warblers; there were thrushes, larks, vieros, finches, sparrows, orioles, bobolinks, blackbirds, blue jays, cuckoos, and mocking birds, with hosts of others more or less musical. Of birds remarkable for plumage there were many fine species. The great blue, white, and snowy herons, and some of the ducks, were very handsome. The snowy owl, well named, was a choice specimen, while red birds, yellow birds, blue birds, scarlet birds, indigo birds, golden birds, and numerous party-colored birds, lent animation to the woods.

Besides all these, the humming bird, bright flashing gem of the air, bred at Cape May. Since the advent of white men upon the coast some varieties of birds have almost or quite disappeared, yet no locality in the United States surpasses Sea Grove and its vicinity in advantages for the naturalist. The distinguished American ornithologist, Wilson, resided during different seasons in the neighborhood of Cape May. At such times he was the guest of the elder Thomas Beesley, of Beesley's Point, and his visits are yet remembered by some of the oldest people. Thomas Beesley declares, in a too brief note to one of his scientific contributions, that the interest awakened there by Wilson in the study of ornithology has never ceased. To that interest and a lively intelligence are to be credited the catalogue of birds and beasts which Thomas Beesley has added to the natural science of his native county, and the fact that "Beesley's Point" has become one of the important centres of scientific interest in South Jersey. It is a legend that birds choose for their habitat the most favorable and pleasant lands—the fairest scenes. Upon this point Thomas Beesley, in a note to his catalogue of Cape May birds, quotes a citizen of Cape May as saying, "If birds in their choice of a residence are gifted in determining what is the fairest and what is best, there can be no question but that the County of Cape May is among the most attractive portions of the earth; for here they congregate in as great a variety and abundance as upon any other portion of at least the civilized globe."

The intermediate latitude of Cape May and its consequent equable climate, with an uncommon distribution of ocean, sound, lake, river, swamp, thicket, wood, marsh, and meadow, afford varied attractions to the denizens of the air. Birds of the north and of the south, with many a feathered beauty "to the manor born," there congregate and dwell, or visit the scene on flashing wing with tumultuous song one after another, as the passing year rolls its changing glories through the sky. To the plodding pot-hunter the birds of Cape May supply—his dinner; to the sportsman, choice and abundant game; to the naturalist, an unequaled field of study; to the artist, forms and hues of beauty; to the invalid, cheer and diversion; to all, a song; to the thoughtful and pious soul, most—bright examples of nature's handiwork, a joyous testimony to the universal providence of God!

The pioneers of Cape May were very practical persons, men who would turn away from the finest display of plumage and the sweetest song to capture a good fat goose or pursue the woodland creatures for their skins; hunting for fur-bearing animals in South Jersey over two centuries ago, they could hardly go amiss. The bison or buffalo, the black bear, the panther, the wolf, the catamount, and the deer, were the largest of the wild beasts of Cape May; of the smaller species there were opossums, raccoons, foxes, minks, otters, and, most valuable of all, the beaver. Some half-dozen kinds of squirrels filled the trees, muskrats infested the streams, rabbits were plenty, and the skunks, in bad odor, were numerous, waiting a change of fashion to give value to their handsome pelt. Twenty years ago a half-dozen black bears in an autumn would perchance be killed in the Cape May County swamps, a few deer would also be taken; the beaver is probably extinct; the opossum, the raccoon, the rabbit, the polecat, the squirrel, the otter, and an occasional fox are the remaining animals of Cape May.

The agents of the Dutch patroons gave little attention to the flowers which adorned the lands they bought, yet a botanist would have gathered them with delight. The same causes which make Cape May the resort of the ornithologist and ichthyologist have rendered all South Jersey a vast botanical garden famous on both sides of the Atlantic, some of its plants being peculiar and local. In 1748 and 1749 Peter Kalm, botanist to the King of Sweden, made a collection in South Jersey, the sight of which made Linnæus forget an attack of gout; the *Kalmia*, a species of laurel, was so called by Linnæus in honor of Kalm. As recent authorities in the botany of South Jersey, Maurice Beesley, M.D., Samuel Ashmead, and Mary Treat, of Vineland, have been extensively quoted.

§ The aboriginal Indian was a savage and a pagan; the mistake of most Christian colonists was to consider themselves saints, and the red man a natural devil. The valleys of the Delaware and Schuylkill were inhabited at the time of the Dutch settlements by the tribe of Lenni Lenape, a name which signified "the original people." The Lenni Lenape were divided into Mantaunaks or Delawares, and

Muncees, Munseys, or Mincees; the last lived above the Sankitan, Stankekan, or Sanhickan falls, near Trenton, and toward the Hudson. The Lenni Lenape were a superior tribe; they came from beyond the Mississippi, and conquered their way to the Atlantic. Subsequently, by the terms of a treaty made with the Hodensaunee Konoshion, or Iroquois Confederation, they abandoned war, becoming "women," that is to say, non-combatants, and, as the Indian matrons were, referees and peacemakers. Hendrick Aupaumut, chief of the Muheconuck, Mohican, or Mohegan tribe, of Stockbridge, Massachusetts, in his report of his mission as the embassador of the United States to the Western tribes (Mem. Hist. Soc. Pa., vol. II.), calls the Delawares "Grandfathers," and adds that the British and Five Nations depended upon them to make peace, as "this nation had the greatest influence with the Southern, Western, and Northern nations;" also that the Lenni Lenape, since about 1600, had been grandfathers or "wise ones," to whom the tribes looked as judges in arbitration.

The general traits of American Indians, aside from the usages of war, characterized the Lenni Lenape; one notable habit of theirs was peculiar to such tribes as inhabited the shores of New Jersey and New York, or lived elsewhere near localities like Cape May. The Indians used no salt, but preserved their fish and meats by drying and smoking; at the shore they boiled, strung, and dried clams, which were used to season their insipid fare. The manufacture of this Indian delicacy left behind an immense quantity of shells, those of the common clam, the *Venus mercenaria*, which the Indians called Pequonuck or Quahaug. These shells, in a broken state, are to be found in great heaps on the shores of the sounds and water-courses in the vicinity of Sea Grove. The fragmentary condition of the shells distinguishes the shell heaps of Indian creation from the beds and mounds of shells which owe their origin to natural causes, or to the bivalve-consuming propensities of white men. The Indian resorted to the shore of the Atlantic, not alone for health and comfort, but to *make money*. Near Sea Grove, as on the shores of Sewan-hacky (Sewan-land), Long Island, New York, an aboriginal "mint" was kept in operation, and the circulating medium of exchange there issued was current at a fixed value all over the continent. This Indian money was called variously sewan, suckauhock, wampum, wampompeague, peague, etc., and was coined in the form of beads, from shells, and strung on strings somewhat after the manner of Chinese "cash."

There were two kinds of sewan. The black—"the gold of the Indians"—was made from the black portion of the clam-shells, and called suckauhock. It was rated at double the value of the white, called wampum, which was made from the stem of the periwinkle (*Littorinæ*); hence the shell heaps the Indians have left along the shore of Cape May contain mostly the white part of clam-shells, broken in small pieces to secure the black and valuable portions. Aside from the color of the wampum, it was criticised by the natives as to its form and finish, and the usages of aboriginal commerce required that the beads should be uniform in size and shape, and bored in the centre. To test sewan, the Indians drew the strings of beads deftly across their noses; if they found them smooth, uniform, and well strung, they passed at par; the worn or imperfect were discounted or rejected. The sewan was used not only as currency, but as jewelry and material for ornamentation. "The Dutch, at Albany," says Kalm, "made and sold a great deal of sewan in their extensive trade with the Five Nations. There were at one time sixty or seventy shops in Albany where sewan was made, and the Iroquois called the town Laaphanachking,—*i.e.*, "the place of stringing wampum." Sewan was also made in other places, "by poor people," and the Indians suffered the inconvenience of "an inflated currency" after a time. The New Netherlands accepted sewan in trade themselves, good wampum being in some colonies as current as silver; it was voted "to goe six a penny in New Haven in 1640." Sewan, or wampum, was the currency of New Netherlands in 1641; afterwards the contributions to the churches were paid in it. At New Amsterdam "four beads of good black, well-strung wampum, or eight of the white," were reckoned as one *stuyver*,—a Dutch coin about a cent in value. In 1650, "there being at present no other specie," sewan was made lawfully current, at the rate of three black or six white beads of "commercial sewan," or four black and six white of the "base strung," for one stuyver, the rate ordered "to goe" in Nieu Haven. By this the drain of "specie" into New England was checked.

The Indian had no banks, and was innocent of "corners," "bonuses," "divvies," brokerages, commissions, margins, "puts and calls," and "irregularities," yet he was a financier in his way, and managed "exchange" for his own benefit. In heavy transactions, sewan, either suckauhock or wampum, was counted by the fathom, measured by the spread arms of an Indian. Commissary Hudde, of Fort Nassau, in 1648, complained that the Cape May tribe made barter "rather too much against them," as "the Indians always take the largest and tallest among them to trade with us," by which means the long-armed "tellers" compassed a long price for their clansmen's beaver-skins.

"In 1756," says Dr. Beesley, "Jacob Spicer, of Cape May, advertised to barter goods for all kinds of produce and commodities, and, among the rest, particularly designated wampum (suckauhock). He offered a reward of five pounds to the person that should manufacture the most wampum. He succeeded in procuring a quantity of the wampum, and, before sending it off to Albany and a market, weighed a shot-bag full of silver coin, and the same shot-bag full of wampum, and found the latter (by weight) most valuable by ten per cent." After the fall of Oswego he chronicles the decline of the wampum traffic. The Narragansetts and Pequods, who were able to produce sewan on their shores, kept themselves rich and powerful by the possession and use of it. The Cape May Indians held similar advantages, and the accumulated refuse of their work shows that they were not neglectful of their opportunities.

Such, two and a half centuries ago, were the people, such the surroundings, among which lay the assumed territories of the High and Mighty Dutch West India Company, and the intended manors of the would-be patroons, Godyn and Blommaert.

§ The unfair advantage Godyn, Blommaert, and a few others had conspired to take of the Charter of Privileges and Exemptions gave great offense, and partisan feeling became bitter against the patroons and those who defended their claims. Director-General Minuit, who was cognizant of the operations of the patroons, was recalled from his office, but Minuit had simply carried out the laws and orders of the company. Sensible of the injustice done him, Minuit transferred his authority to the Manhattan Council, and sailed for Holland to vindicate himself, in March, 1632, bearing with him not only his own troubles, but sad news for the patroons and the friends of the colonists at Swaanendael.

The first accounts from Swaanendael received by the patroons, some time after the Walvis left that colony, reported that all had been well, and that the colony was pleasantly prosperous. The ill luck of the Walvis had discouraged the proprietors somewhat, but Godyn was still sanguine about the whale fishery, and, in February, 1632, it was agreed that a ship and yacht should be fitted out, with De Vries himself as patroon and commander, to fish in the South Bay during the winter of 1633. This ship and the yacht Squirrel were accordingly fitted out for a whaling voyage, and were ready to sail the last of May. On the 24th of May, just before De Vries got off, news was received at Amsterdam, having been brought by Director-General Minuit, by the way of Portsmouth, that Swaanendael had been destroyed by the Indians.

De Vries, though distressed by the news, put to sea, but an unskillful pilot ran his ship on the sands off Dunkirk; she with difficulty got into Portsmouth the 25th of May. The ship was made seaworthy, and sailed the 1st of August, in company with the great ship "New Netherland, of six or eight hundred tunnes," which had been built at Manhattan, in 1631, and was then returning from her first voyage to Holland.

De Vries arrived on December 5th, in the offing of Godyn's Baai. As he neared the coast he saw no beacon kindled to give warning of his approach; he heard no resounding and reassuring gun; no signal waved to denote his looked-for arrival, and give the sign for joyous welcome. An ominous silence brooded everywhere,—only the waves dashed mournfully, and the tall cedars soughed in the blast of December, as if they chanted a requiem. No Indians appearing, a well-armed boat was sent into the Horekill the next day, to open communication. Finding none of the savages about, the boat pushed on, and landed at Swaanendael, where discoveries were soon made which justified the worst apprehensions of De Vries. The colony had disappeared,—buildings, gardens, plantations, fishing-stations, whale-boats, all were gone. Only ashes and fire-blasted ruins remained, surrounded by the wolf-gnawed and bleaching bones of his comrades and servants.

In despondency De Vries returned to his yacht, and a gun was fired to call in the Indians. The next morning a smoke was seen arising from near the ruins of Swaanendael. The boat went into the creek, and a few of the savages were seen prowling about. They were shy, and the crew of the boat distrustful. The yacht gave more protection from treacherous arrows than the open boat, and so De Vries ran her into the creek. The Indians soon came to the shore, but for some time none could be persuaded to come on board. Finally one venturesome fellow made bold to dare the vengeance of the *Swannekins*, and came alone among the Dutch. De Vries gave him a "cloth dress," and sent word by him to his chief that he wished to make a peace. The Indians at once became more familiar, and that night one of them stayed on board, and was induced to give the particulars of the tragic fate of the colony.

According to the story of the Indian on the yacht, Gillis Hossett had considered it requisite to post the arms of Holland, painted on a sheet of tin, by attaching them to a pillar he set up, the site of which the Indian pointed out. An Indian, attracted by the sheen of the metal, "not thinking he was doing amiss," carefully removed the shield for his own purpose. Hossett took much to heart the insult to the Batavian Republic, and angrily denounced the tribe for the offense of a person, as if it were some mighty matter. It was a great fuss to make about a bit of tin, but the Indians took it for earnest, and soon presented Hossett the scalp of the culprit, to his avowed astonishment, chagrin, and disgust.

Rebuked, humbled, thrown off, hurt in feeling, the jealous, vindictive sons of the forest returned to their wigwams, but not to live in peace. The Indians had a custom like that of the Jews, in "the avenger of blood." If a relative were slain, it was an obligation to avenge his fall unless "atonement" were made by the offender. This could be done by his paying, after the manner of the ancient Greeks, "blood-money," to "cover the graves of the dead."

"If a brother bleed,
On just atonement we remit the deed;
A sire the slaughter of his son forgives,
The price of blood discharged, the murderer lives."
(POPE: *Iliad*, ix.)

The Indian who had been killed at Swaanendael was a sachem,—vengeance could not be allowed to sleep. The aggrieved Indians held Hossett accountable as the cause of the murder, still he could at any time have purchased exemption for a few guilders' worth of goods; this he unwisely neglected to do, and was accordingly condemned to die, and the colony that harbored him was to share his fate.

One day Hossett was sick and remained in the house, but one of his men, a housekeeper, being with him, when a lurking war party of Indians came near the place. In the yard a large bull-dog, or Dutch mastiff, was chained; had he been loose, they would not have dared approach the house. Suddenly three Indians presented themselves, and offered a small lot of beaver-skins for sale. Learning that no others were near, they set upon Hossett and his servant and killed them at once. With the dog, "which they feared most," they had more trouble, and the Indian related with wonder and admiration that the brave guardian of the threshold never ceased to fight, and died only when pierced by twenty-five arrows. But for his chain, as they knew, the Dutch mastiff would have taught the bloody savages the difference between a dog of his breed and keep and one of their own skulking, mangy little curs. The men of the colony were at work in the adjacent gardens and cornfields; they were approached in a friendly manner, and a treacherous attack made upon them. Whatever of courage they manifested, whatever of desperate heroism (for the Dutch were brave), is unknown, as it was unavailing; one by one rapidly they fell, far from their beloved "Faderlandt," among barbarous foes, perishing victims to the folly of their Governor and the revengeful passions of cruel savages.

Shocked, saddened, disappointed, and involved in financial loss, De Vries was not discouraged, and made no useless attempt at revenge. The Indians were glad to make a formal treaty of peace with De Vries, which was brought about by his tact and coolness the following day. Receiving various presents, the bewildered Nanticokes departed in great joy to hunt for beaver-skins to trade with the prudent and reticent Hollander. Such is the awful story of the first bloodshed in the settlement of Delaware, and thus were the possessions of the Dutch "sealed with blood, and dearly enough bought." To De Vries the honor is due, that from that time war between the races was unknown, and bloodshed extremely rare in all the country round about Swaanendael.

Mindful of the plans and interests of his partners, De Vries tried the whale fishery; he had anticipated "royal work," but, from the imperfection of their gear, the Dutchmen were not very successful. To eke out his supplies, De Vries, in his yacht, the Squirrel, with seven men made a trip up the Poutaxit, as the Indians called Zuydt, South, Godyn's, or Delaware Bay; and above into the Lenape-ittuck, Mackerish-kitton, or Arasapha, as the red men had named the Zuydt, South, Godyn's, Prince Hendrick's, or Delaware River. It was New Year time, and the Dutch hoped to "buy some beans of the Indians." The bay and river were full of floating ice; working his yacht through this, De Vries came, on the 5th of January, to Fort Nassau, finding none but Indians.

The natives advised De Vries to go up the Timmer Kill, or Timber Kill, for his supplies; but a Sankitan, or Stankekan squaw warned the Dutch to keep out of the creeks, or the river Indians would murder them, as they had recently killed the crew of an English shallop, in "Count Ernest's River." Avoiding the creeks, therefore, De Vries went on up to Red Hook, or Mantes. There some forty Indians came on board, offering to barter beaver skins, and "playing on reeds to allay suspicion." Unaware that the Dutch were informed of their murder of the English crew, some of them wore the jackets of the men they had butchered. De Vries told them their "Maneto" had revealed their treacherous plans to him, and, driving them all on shore, returned to Fort Nassau. There several chiefs came on board the yacht, some of whom had worn the jackets at Red Hook, but now they were dressed in robes of fur. The Indians sat down in a solemn circle on the deck, and stated they had come to make a long peace; a long ceremony, during which ten beaver-skins were presented, one after another, by the Indians, ratified the formal compact. For the skins presented in their ceremonies the Indians refused any compensation whatever. De Vries, however, bought other beaver-skins, and, procuring a small supply of corn and beans, sailed for his ship, and was on board the 13th of the month.

Five days after, De Vries again started to coast along the shore and visit Fort Nassau. On the way he was a fortnight frozen into "Vineyard Creek," where the Dutch shot a multitude of turkeys, "weighing from thirty to thirty-six pounds" each. It was the 3d of February before the yacht could be got up to its destination. By that time a war had broken out between the Minquas and the Sankitans, and no corn could be had. After much trouble from the ice, the yacht was got back to the ship, where a joyous welcome was given the long absent and adventurous voyagers by their anxious shipmates.

Still short of provisions, and ambitious to be the first Hollander to visit the Chesapeake, De Vries sailed on the 5th of March for Virginia. He visited Sir John Harvey, Governor of Virginia, and was courteously entertained by that noble knight. De Vries made Governor Harvey acquainted with the Dutch operations on the Delaware, and was able to identify the English crew whose murder he had heard of at Fort Nassau as one of eight men which Governor Harvey had sent the previous September into the Delaware, in a sloop, "to see if there was a river there." The Governor imagined his men "to have been swallowed up in the sea," having heard nothing from them until the sad news by De Vries made him acquainted with their fate. The patroon of Swaanendael and Cape May remained a week at Jamestown, and then, with an abundant supply of provisions, and a present of goats for Manhattan, where Governor Harvey had heard there were none, he returned to his fishermen in Zuydt Baai.

Once more warmly welcomed by his company, the patroon learned that several whales had been captured, but more lost after being struck, the harpoons being defective. De Vries had, however, become satisfied that the whale fishery was less profitable than the fur trade, and proposed to carry out his original intention of a voyage to Newfoundland and the Saint Lawrence for fish and peltries. Wishing to examine the coast, De Vries sailed on board his yacht for Manhattan the 14th of April, and, coasting northward for two days, arrived safely at Fort Amsterdam, leaving Swaanendael, Godyn's Baai, and the Arasapha once more to the whales, the savages, and the aboriginal wildness of nature.

§ When De Vries arrived in New Amsterdam, on the 16th of April, 1633, he found the new Director-General, Wouter Van Twiller, on board the ship Soutberg, which had just arrived in the harbor. The information which De Vries gave Van Twiller aroused him to take measures to hold possession of Zuydt Baai, and the fur trade in the country adjoining; accordingly Arendt Corssen was appointed commissary, and instructed to purchase a tract of land on the Schuylkill for a plantation and trading post, for both of which purposes the location there was highly esteemed. Corssen bought "for certain cargoes," from "the right owners and Indian chiefs," a tract called "Armenveruis," lying about and on the Schuylkill. The Indian title being thus secured, the Dutch took formal possession of Pennsylvania, and established a trading house there, which, though soon abandoned for a time, was afterwards enlarged to a post or station and called Beversrede, being situated within the present bounds of Philadelphia.

Among the improvements ordered by Van Twiller for the year 1633 was "one large house," to be built at Fort Nassau on the Delaware. The work must have been neglected, for in 1635 a small party of English from Point Comfort, Va., under the leadership of Captain George Holmes, and, as some have said, in the interest of Sir Edmund Plowden and his associates, took possession of Fort Nassau, which they found vacant. Thomas Hall, one of Holmes's men, deserted at Fort Nassau, and, reaching Manhattan, gave information to Van Twiller. A Dutch force soon captured Holmes and his party, and took them to Fort Amsterdam, from whence they were sent, "pack and baggage," back to Virginia. The Dutch, after the affair with Holmes, repaired and garrisoned Fort Nassau, and gave more attention to the valley of the Delaware. The administration of Wouter Van Twiller ended early in the spring of 1638, he being superseded by William Kieft.

§ Although the Dutch were the first to "occupy" the Delaware, Governor Sir John Harvey, of Virginia, had sent an unfortunate expedition there in 1632; and before the patent for Maryland was sealed that year, Sir John Lawrence, Sir Edmund Plowden, and others applied to King Charles I., of England, for a grant of Long Island and thirty miles square on the mainland, which they proposed to call "Syon." After the death of the first Lord Baltimore, before the full execution of the formalities of the grant of Maryland made to him, Plowden and his associates made a second application; this time asking for Long Island, and the small isles between the thirtieth and fortieth degrees of north latitude, within six leagues from the mainland near Delaware Bay, and forty leagues square of the adjoining coast; to be held as a County Palatine, and called New Albion, "with the privileges as heretofore granted to Sir George Calvert, late Lord Baltimore, in Newfoundland." The king confirmed the grant made Lord Baltimore to his son and heir, but one month after the sealing of the Maryland patent the king (says Neill), on July 24th, 1632, ordered Sir John Coke to issue a patent for Long Island and the adjacent country to Plowden and his associates.

On the 23d of September, 1633, Captain Thomas Young, gentleman, received a special commission from the King of England to organize an expedition and explore in America. This expedition sailed in the spring of 1634, and with it came Master Robert Evelyn, Captain Young's nephew, as lieutenant. The voyage of Captain Young was in connection with the enterprise of Plowden for the settlement of New Albion, but from stress of weather, or lack of a pilot, his course led him into the Chesapeake; there desertion weakened the company, but Young fitted out a shallop or pinnace at Jamestown, in July, and sailing to the Delaware with about fifteen men, established as the headquarters of New Albion a post he called Eriwomeck, near the mouth of the Schuylkill, at Fort Beversrede, which the Dutch had just abandoned. In September, as has been noted, George Holmes seized for the New Albion Company the vacant Fort Nassau, from which he was soon ousted by the Dutch. Lieutenant or "Master" Robert Evelyn went to England, early in 1635, upon some errand from which he soon returned; in 1637, he was appointed a surveyor by the Governor and Council of Virginia, but missed confirmation; he was afterwards proxy for St. George's Hundred, in the Maryland Assembly, but was again in England in 1641. At that time Evelyn and others published "a card," describing the valley of the Delaware as a fine place, where the English had traded since 1527, and where Evelyn himself had been stationed for four years with fifteen men, trading and exploring in safety. On Evelyn's return from England he was commissioned, June 23d, 1642, to command and drill the militia, at Piscataway, four miles below Washington. The identity and character of Evelyn are

LAKE LILY, FROM THE NORTH.

important in this history, as he was the first recorded explorer and geographical describer of Sea Grove and Cape May, as is elsewhere related.

Captain Young continued his explorations about the Delaware for about eighteen months, hoping to find there the entrance of a passage to India; he became satisfied of the importance of the inland or back country, and, in 1636 or the year after, returned to England and asked for himself and his company a grant of whatever inland regions he might discover and explore. Sir Edmund Plowden remained in America until 1648, trying to settle the territories of New Albion, of which he was Earl Palatine; he did not succeed in this, owing to the opposition of the Dutch and others along the Delaware. Having exhausted his fortune during his stay of "about seven years" in this country, Plowden returned to England, by the way of New Amsterdam and Boston, "for supply." In London, in 1648, under the name of Beauchamp Plantagenet, Plowden published his "Description of New Albion," an inaccurate pamphlet, a copy of which remains in the Philadelphia Library.

Beyond elaborating and publishing a remarkably liberal, just, and worthy plan of government, the enterprise of the New Albion associates achieved nothing of note. Sir Edmund Plowden himself was the descendant of an eminent jurist; he was as unhappy in domestic life as unfortunate in business; his wife Mabel, daughter of Peter Mariner, of Wanstead, Hampshire, England, left him after a married life of twenty-five years, alleging abuse as her cause. Sir Edmund came to Virginia, and was at Eriwomeck, as Earl Palatine of Albion, in 1642,—"the fort given over by Captain Young and Master Evelyn." He was visited in London by some Marylanders in 1652, but he never left England again. Made poor by his outlay in behalf of his scheme of colonization, Plowden's fortunes became desperate; he was arrested for debt, and died in the debtors' prison in 1655. There is a pathos about the fate of the earnest Palatine of New Albion, which is made more effective by a statement of the social ideas by which he and his associates proposed to be governed.

The pioneers of New Albion raised less tobacco and sold less rum for beaver-skins than their neighbors, but they were the first to comprehend the vast width of the continent; and in evidence of their culture and character, they presented the world with an illustrious example of political sagacity in a model form of free and liberal government. While kings and ecclesiastics conspired in Europe to enslave the bodies and the souls of men, while Boston and New Haven fostered despotism, and called it theocracy, Roger Williams, dividing his land with all who needed, founded a state purely on the will of the majority, with God alone as the Ruler of Conscience; and Sir Edmund Plowden, beside the Delaware, sought to establish a more liberal, wise, and perfect organization of society than the world had ever known. Rhode Island became a more complete "Democracie," and fortunate Connecticut grew to love freedom by experience, but New Albion formulated the principles of political order, and put forward her ideal proposition, at once and entire. Of little consequence now are the "Manors, dignified by well chosen names, giving titles to each of the Earl's family"; of less account, the "Albion Knights of the Conversion of the Twenty-three Kings"; less still the mere ghost of an established church, barely provided for in a document which might have been quoted as the death warrant of state religions!

Guarding against demagogue usurpation, the institution of New Albion enfranchised the people, and deferred to popular intelligence; obedient to British usages, it still insisted upon independence and freedom, and thereto obtained the sanction of the throne. Mildness, humanity, and justice were characteristics of the whole constitution of the intended state, and, most glorious of all, entire religious freedom was guaranteed; dissent was not amenable for punishment, and heresy to be proceeded against only by education; with the proviso, that "*this argument or persuasion in religion, ceremonies, or church discipline, should be acted in mildness, love, charity, and gentle language!*"

§ As early as 1626, Gustavus Adolphus, the illustrious King of Sweden, the champion of Protestantism in his time, undertook to found a Swedish colony on the shores of the Delaware; this was first suggested to him by the same William Usselinx, of Holland, who in 1590 proposed the Dutch West India Company to his countrymen. Usselinx waited upon Gustavus, and being a learned man, unusually well informed upon matters in America, he convinced the king and his nobles of the desirability of a Swedish-American colony, and of the feasibility of a great Swedish trading corporation to establish such a province.

The company was organized duly, "to trade to Asia, Africa, and the Straits of Magellan," and on July 2d, 1626, the king issued an edict at Stockholm, "in which he offered to people of all conditions liberty of shares by subscription, according to their ability or inclinations. The proposal was received with general satisfaction," say the "Annals of the Swedes." Gustavus took for himself stock to the amount of four hundred thousand dollars, at equal risk. The king's mother, and Prince John Cassimir, his brother-in-law; the members of his majesty's council; many civil and military officers of high rank; the bishops and other clergymen; many merchants and citizens; country gentlemen and farmers, became subscribers; ships were fitted out, and all requisites for trade and a colony provided; an admiral, vice-admiral, commissioners, merchants, and other proper persons were appointed, and a few vessels started for America.

The Swedish cannon were the speakers and champions whose eloquence clinched the arguments of the Lutheran Reformation, and recast the destinies of a thousand years; but for Gustavus and his guns, the Protestant movement would have ended in the beginning. This hero king and philanthropist, whose mind was as practical as it was comprehensive and brilliant, who in defense of religious freedom invented and victoriously used modern artillery, was not inclined to hesitate in an enterprise which he declared to be "for the benefit of the persecuted," for the security of "the honor of the wives and daughters" of those made fugitives by war and bigotry, for "the good of the common man," for the blessing of "the whole Protestant world," and "the advantage of all oppressed Christendom," through undue deference to the dubious and conflicting claims of ambitious potentates, or the greed and avarice of monopolizing corporations.

Neither was the King of Sweden careful, like the States General of Holland, to avoid direct responsibility for colonies. "Every inch a king," after the best manner of his times, his charter declared that "politics lie beyond the profession of merchants," and reserved the government of all future Swedish colonies to a Royal Council. Thus the formidable cannon of Sweden and the invincible sword of Gustavus were pledged to the protection of the emigrant. The privileges of the Swedish Company were open to all, and colonists were invited from every nation of Europe; slaves were discarded, as a laborious and intelligent Swedish population, with wives and children, it was wisely thought, would be quite as profitable and more to the honor of the state. In Sweden all was in readiness for the colony, when, through the influence of the papal power, war was provoked, and broke forth suddenly. Gustavus found himself compelled to invade Germany "to vindicate the rights of conscience," establish toleration, and secure German liberty by defending the principles of the Reformation. The fight was for the safety of Protestant Christendom. In the emergency the funds of the new trading company were, as a military necessity, diverted for a time to the purposes of war; yet the king abated not at all his zeal for the American enterprise even on the field and in camp, and from Nuremberg, October 16th, 1632, he communicated to Oxenstiern, his great minister, enlarged and most liberal plans for the proper setting of that "jewel of his kingdom," even in case of his death.

At the battle of Lutzen, November 3d, 1632, Gustavus fell. His death changed the course of European politics; the project of Swedish colonies was temporarily postponed. The little squadron which left Sweden for America, perhaps on private account, about the time of the formation of the Swedish Commercial Company, as Swedish ships may have done before, was attacked at sea by Spaniards, some of the ships being captured; but there is reason to believe that others escaped and reached the Delaware, where their factors engaged in trade with the

Indians, and that from that time there were always a few Swedes and Finns in the valley, who finally located among the aborigines, well up the river.

Peter Minuit, finding partisan influence too strong for justice in the Dutch West India Company, visited Oxenstiern and the court of Sweden, and offered to conduct a Swedish colony to the unoccupied west shore of the Delaware. In the spring of 1638, a man-of-war, named the Key of Calmar, and a tender, the Griffin, from Gottenburg, Sweden, with about fifty emigrants, under command of Peter Minuit, as Governor by commission of Queen Christina, put in at Jamestown, Virginia, to "refresh with wood and water," being bound for Delaware Bay, which is the confines of Virginia and "New England," "to make a plantation." The Treasurer of Virginia desired to have a copy of Minuit's commission, but the Swedish Governor declined to show his charter, unless he could arrange for the purchase of tobacco for shipment to Sweden; but the colonial laws of England did not permit such a traffic, and so Minuit, after spending ten days in the Chesapeake, pursued his voyage, and entered "Zuydt Baai" early in April. The Swedes soon after disembarked at Missipillion Point, twenty miles up the bay, on the western shore.

Emigrating from an almost arctic climate, the Swedes were delighted by the Eden-like airs which, in April, are the atmosphere of the capes of Delaware, and which linger over them through the balmy summers. Enchanted with the climate, and charmed by the scene, they gave their landing-place the name of "Point Paradise." It may have been an extravagant appellation; but as the lover of natural beauty sits quietly at Sea Grove, and sees the glorious summer sun sink amid his clouds, in the waters of the bay, above Henlopen and far-away Missipillion, he need not be a poet to imagine that the scene is somewhat too fair to be all of earth.

§ When, in 1623, the French attempted to take possession of the Delaware, they were prevented "by the Dutch settlers there;" so, in 1635, the Dutch from Manhattan ousted the party of English under Holmes. In 1638 they were in the river, and equally ready to repel the Swedes. Soon after the Swedes arrived in the Delaware they were visited by some officials of the Dutch West India Company, who notified them of the claims of Holland thereabouts, and warned them out of the bay. The Swedes, in answer to this challenge, stated that they were on their way to one of the West India islands, and had put into Zuydt Baai but for refreshment after a prolonged and stormy voyage, which they should continue as soon as they supplied themselves with fresh meat, water, wood, and a few necessaries. Not inhospitable, the Dutch consented to this delay, trusting to the representations which had been made to them.

But Minuit, who well understood the Dutch policy and the extent of their jurisdiction, merely moved his expedition up stream, beyond the limits of that which had been Godyn and Blommaert's purchase in 1629, and at Paghacking, or Minquas, Creek, near Wilmington, Delaware, made a second landing. There, from a local sachem, named Matteehoorn, a plantation was bought, "between six trees," "a kettle and a few trifles" being paid in consideration. The Swedes won Matteehoorn by the promise of the half of a crop of tobacco, to be raised on the ground he conveyed to them. Between Matteehoorn and the Swedes some document, memorandum, or deed was drawn up; "as no Swede could yet interpret Indian," and no Indian understand Swedish, the paper "was written in low Dutch." The Indians could *read* neither language, and were, it seems, induced to sign a deed of the land from Cape Henlopen to Trenton, or "Sankekan" Falls, and as far inland as the Swedes might gradually require, under the impression they were conveying a mere patch of ground to raise tobacco on "at halves." The Swedes, says Indian tradition, never divided the tobacco, but held the Indians to the letter of the fraudulent deed.

The mouth of the Paghacking was but twenty-five miles from the Dutch Fort Nassau, and messengers were soon sent to learn Minuit's intentions; these he cajoled with courtesy and fine words, and they went back to their fort. In a few days the people of Fort Nassau came down again, and found the Swedes "had done more,"—buildings were begun, goods disembarked, and a small garden made. The Dutch asking what it meant, Minuit made various excuses and pretenses, still declaring his intention to soon depart. As soon as the Swedish colony was safely established, Minuit revealed his purpose by sending his small vessel, the Griffin, up the river for Indian trade. She was not allowed to pass Fort Nassau, and Peter May, the sub-commissary, boarded her and demanded her commission. The Swedish master refused to show his papers, and defended the establishment of a Swedish colony on the Delaware, saying his queen had as good a right to build a fort there as the Dutch West India Company. Of all this the people at Fort Nassau took note, and at once forwarded the particulars to Manhattan. Director-General Kieft protested against the Swedish colony, and warned them to depart at once, as all that part of the world, especlally the Delaware, belonged to the Hollanders, it having been for a long time "beset with forts and sealed with the blood of the Dutch." But the epistle had little influence with Minuit, and Kieft, who was as "economical" as he was "testy," was too prudent to attack the colony of a nation as gallant and victorious as the Swedes.

There were but fifty souls in the first expedition under Minuit, and of these many were "bandits," condemned to penal servitude. Yet, notwithstanding the opposition from the Hollanders, the little colony "between six trees" was prosperous. On the north bank of the Paghacking, two miles from the Delaware, a fort was erected, and the name of Christina given to it and the creek,—the arms of Sweden being carved with the royal monogram on the boundary posts of the station. Besides, a plantation was made, where corn, beans, squashes, and the profitable tobacco, grew as they long had grown in the same region, except that they showed, by their unexampled productiveness, the difference between the bone paddle of the overtaxed squaw and the heavy steel mattock of the athletic Scandinavian. Not only were all the Indian products improved by Swedish culture, but the seeds of Europe were introduced, and soon made evident, by prolific increase, the proverbial fertility of the soil of Delaware and the influence of a genial climate. Meantime, commerce was not neglected; the goods brought for barter were soon disposed of, for the Swedes undersold the Dutch, and it is recorded that the beaver-skins taken to Sweden the first year of Minuit's administration damaged the Dutch trade on the Delaware more than thirty thousand guilders.

About midsummer the vessels which brought the colony returned to Sweden, but Minuit and twenty-four men, with a good supply of merchandise and provisions, remained at Fort Christina. There was great delay in the coming of further supplies from Sweden, and in the winter of 1640 the Swedes were so much in want they decided to abandon their plantation, and merge themselves in the settlement at Manhattan. But early in the spring, the day before the Swedes had decided to give up Fort Christina, a ship named the Fredenburg, Captain Jacob Powelson, of Utrecht, Holland, arrived with a company of Hollanders, who, under the Dutchman Joost De Bogaredt as a commander for Sweden, had been sent out by Henry Hockhammer, according to grant and agreement with the Swedish Government, to settle as Swedish colonists on the Delaware. The distress of the resident Swedes was relieved by De Bogaredt, and they continued at Fort Christina. Another colony was begun a few miles below, and soon the trade of the Dutch West India Company on the South River was "entirely ruined."

In the fall of 1640 Peter Hollendare came from Gottenburg to Fort Christina as Deputy Governor of the Swedes in America; two vessels soon followed, and "a new treaty was made with the Indians for more land." The Swedes called their territory Nya Swerige, or New Sweden, and to Zuydt Riviere they gave the title of New Swedeland Stream. Nye Swerige was more fortunate than Swaanendael; it had become a successful colony,—the first *permanent* settlement on the Delaware. Minuit had proved a good guide and a sagacious, even if crafty, commandant; but his work was done. In 1641, according to Acrelius, the Swedish historian, he died at Fort Christina, while Peter Hollendare continued the government.

The colonists of Virginia, as early as 1629, extended by Nathaniel Basse an invitation to such of the people of New England as preferred a fertile soil and mild climate, to come and settle in the Valley of the

Delaware. The matter was discussed among the Puritans, but the first adventurers sent to the Delaware by them were from New Haven, in 1638, the year that colony was founded. The traders of New Haven, George Lamberton and others, led the way. The project of emigration was originated by a few enterprising persons, who soon formed a company that finally sold out its interest to the community at large, which, as a Church, desired to establish a mission among the Delawares, and found a prosperous colony where all should live in godly order, and their children after them "should continue to abide under the wings of Christ."

Captain Nathaniel Turner bought of the Indians, for £30, the land along shore from Cape May to Raccoon Creek, Varcken's Kill, Hog Creek, or Salem River; the deed was dated November 24th, 1638. At different times during the next two years additional lands were purchased by and for the New Haven adventurers. They were helped in their negotiations by a refugee sachem of the Pequods, and represented that their lands cost them £600 in all. (*N. H. Col. Rec.*)

In April, 1641, an expedition of some twenty families, or sixty or more persons, sailed for the Delaware in Lamberton's bark, or ketch, under command of Robert Cogswell. Voyaging by the way of Manhattan, they were detained by Kieft; but promising allegiance to the Dutch if they settled in Dutch territories, they were allowed to go on. The New Haven people landed on Varcken's Kill, near Salem, New Jersey, and "on the Schuylkill." Trading houses and habitations were erected on Varcken's Kill. The Schuylkill settlement was at or near Fort Eriwomeck, the headquarters of New Albion; the Dutch "Beversrede," the Indian Armenveruis, or Passyunk, at Philadelphia. These plantations were to be governed "in combination" with New Haven, and Captain Turner was furloughed from New Haven and authorized to go to the Delaware, "for his own advantage, and the public good in settling the affairs thereof."

Though the New Haven people were intruding upon territories claimed by both the Dutch and Swedes, yet such was the confusion of titles that their claim may have been supposed by them as good as any; besides, they found Sir Edmund Plowden in the bay, with an English grant of New Albion, and gave allegiance to him as Earl Palatine. Kieft, however, considered that Cogswell had purposely deceived him, and the Swedes were ready, as they had agreed, to cooperate to "keep out the English." In May, 1642, two sloops, the Real and Saint Martin, with thirty men, under Jan Jansen Van Ilpendam, of Fort Nassau, were sent by Kieft's orders to break up the English settlements on the Delaware. Fort Eriwomeck was first visited; there were some Marylanders at that place of the rougher sort, and accounts differ as to the result of Kieft's proclamation, which was read to them. One author asserts that the English were so violently blasphemous and threatening, that Jansen drew off his sloops and made the best of his way out of the Schuylkill; but others declare the colony there was broken up.

From the Schuylkill Jansen sailed to Varcken's Kill. There, meeting no resistance, he burned the English buildings, took possession of all the goods, and bore away most of the people as prisoners to Fort Amsterdam, at Manhattan. Their goods were restored to the New Haven people, and they proceeded home to Connecticut. The colony on Varcken's Kill had been very unfortunate; the members had come on foot from Boston to New Haven, where they remained but a short time before moving to the Delaware; the winter they spent on Varcken's Kill was excessively cold, and the summer had been very sickly; their time, their trouble, the cost of their lands, all were lost, as well as damage done their goods. Still the undaunted Lamberton continued to trade in the Delaware from New Haven, though annoyed and interrupted at times; the New Haven people also attempted, though in vain, to renew their colony, being turned back at Manhattan. The records of New Haven for a few years show the public and private loss from the Delaware enterprise. The sufferers applied to the Commissioners of New England, to Oliver Cromwell, to Richard, his son, and finally, to Nichols,—when he first came out,—for restitution at the expense of the Dutch. Their losses were more than a thousand pounds sterling, but, from one cause and another, nothing was ever realized by them in return.

There was no original and permanent colony from New England on the Delaware until the whalemen, who first appear on record in 1685, settled at Cape May. Although Plowden, who never had many men with him, had been unable to defend his earldom, or protect the people who recognized him as their lord, and although the colony was driven out to return no more, still members of the Calvinistic community were left behind, and the fame of the Delaware was spread abroad by the quarrels which followed. In the settlements of the following generation around the bay, the Yanokies (silent men), as the Mais Tchusaeg, or Massachusetts Indians, called the New Englanders, had their full share of action, influence, and honor, as is usual everywhere. Comparing a record of the early settlers of New Haven and Cape May, about one-fifth of the family names from the Cape May list are inscribed on the older New Haven document.

§ Peter Hollendare remained as the successor of Governer Minuit but eighteen months. On the 15th of February, 1643, after Hollendare's return to Sweden, Colonel John Printz arrived at Christina, and at once assumed office by virtue of his commission as Governor for the Queen of Sweden. Governor Printz came out in the ship Fame, attended by the Svan or Stork, and by the Charitas—all armed vessels. The instructions of the new Governor were full and explicit. About a

THE PAVILION. 1875.

hundred soldiers came with him, as well as many colonists, the royal council having appropriated over two million dollars annually for the support of his administration. Printz was directed to keep on good terms with the Dutch and cultivate trade with the English of Virginia, and especially to see that the Indians were treated with consideration and justice, as the original owners of the soil. Still, the Swedes were to assume control of the Delaware, "that the river may be shut," and in case of aggression on the west side they were commanded to "repel force by force;" Printz was thus "to take care" of his jurisdiction. On Tenacong, now Tinicum Island, Printz built the fort, New Gottenburg, of "vast logs," and erected Printz Hall for his residence. To shut up the river, a fort was built on Varcken's Kill, called Helsingborg or Elsingburg; it had three angles, and mounted eight twelve-pound guns.

The Rev. John Campanius, of Stockholm, came with Printz as chaplain; Reorus Torkillus had served in that capacity at Christina from the first. He died the 7th of September following the arrival of his colleague, being but thirty-five years old, still memorable as the first Lutheran missionary in the Delaware Valley, if not in all America.

There is a well-authenticated tradition related by the Swedish botanist, Peter Kalm, in 1748, upon the authority of Moons Keen, one of the ancient Swedes, regarding Fort Helsingborg. When work was begun upon the fort, the builders found traces of ancient occupants in certain wells, which were bricked up to a depth of twenty feet or more under ground; there were vessels and fragments of pottery, with broken and displaced brick also found near by, giving unmistakable evidence of the civilization of former residents. The situation of the wells and the position of the other relics was in a meadow near the river, where all the surroundings indicated the absolute antiquity of the pre-historic settlement. The Indians, who had occupied the ground for generations, had no knowledge or tradition of people who dug wells and used bricks and pottery in a civilized manner, but assured the Swedes the relics had certainly been where they found them for more than a hundred and fifty years,—ever since the voyages of Columbus. Were these wells the work of Lief Erikson, and the Norwegian Christians, A.D. 996 to A.D. 1000? Were they dug by the men who built the round tower at Newport?

§ In October of 1643, the year Helsingborg was established, De Vries again visited South River, putting in as he was on his way to Virginia. As the craft came abreast of Fort Helsingborg, a gun was fired for her to strike her flag and "come to." Blanck, the schipper, asked advice of De Vries. "If it were my ship I should not strike," said De Vries, "for I am a patroon of New Netherland, and the Swedes are mere intruders in our river." The schipper, however, "had a desire to trade," and lowered his colors. A boat came on board the vessel at once, and she sailed up to Tinicum that afternoon. The Dutch were welcomed to Fort Gottenburg by the Governor, who "was named Captain Printz, a man of brave size, who weighed over four hundred pounds." Being informed of the position of De Vries and his doings on the Delaware, Printz drank his health in "a great romer of Rhine wine." The Dutch traded confectionery and Madeira wine for beaver-skins at the fort for five days, and then visited Fort Nassau, where a garrison of Dutchmen was found. Returning to Tinicum, De Vries went with Printz to Fort Christina, "where there were now several houses," and spent the night with the Governor, who "treated him well." On parting from the Swedes the Rotterdam vessel fired a salute in honor of their hospitality, and sailed away for Virginia. Thus De Vries, who forebore his vengeance upon the feeble Nanticokes, and ever counseled justice and peace in dealing with the bow-bearing Indians of Manhattan, was brave enough of himself to defy a battery of cannon in an unarmed vessel, and courteous enough to win the favor of a supposed enemy and competitor.

Patroon De Vries spent the winter of 1643 in Virginia, and sailed from there for Holland, where he arrived in June, 1644. De Vries had given his best efforts for a dozen years to New Netherland, but the petulance of Hossett, the mismanagement of Van Twiller, and the stubborn folly of Kieft, had thwarted his sagacious endeavors, and to him the memory of his sojourn in the New World was a sad retrospect of losses and tragical disappointments; he seems never to have revisited America.

David Pietersen De Vries was one of the finest characters of New Netherland history. A man of the people, he was ever a foe to despotism, injustice, and cruelty. In Manhattan, where he resided so long and honorably, he was, as Chairman of the Citizens' Committee, the acknowledged head of the Dutch democracy. The Indians trusted De Vries as a Swannekin "who never lied like the others," and his influence with the aborigines, with his characteristic tact and discrimination, more than once saved the province from destruction.

To the folly and mismanagement of Van Twiller De Vries opposed the coolness of practical sense and the courage of a hero. When the foolhardy and barbarous Director Kieft ordered the massacre of his Indian refugee guests, De Vries gave earnest warning, and the revengeful ruin which followed came upon Manhattan despite the protest of the democratic leader. Firm and perhaps overbearing in maintaining his own rights as a citizen and privileged proprietor among his equals, even at the cannon's mouth, he forebore revenge upon the ignorant savage trespasser, and ever counseled and practiced honesty and humanity in all dealings with his Indian neighbors. Wise in council, prudent in action, De Vries stood firm for right, palliated the evils he could not avert, and constantly manifested that self-control and magnanimity which won the affection of the Indians from Fort Orange to Sandy Hook, and conciliated the barbarians of Swaanendael and Scheyichbi, making smooth and peaceful the ways of his successors on the Delaware.

Though filling a merely subordinate position, De Vries was by nature and experience equally commendable as a man, a citizen, a commander, a diplomat, or a statesman. It would be untrue to history and unjust both to him and his creed not to record, in addition, the fact that the first resident patroon and owner of Cape May was a man of religious sentiments, in principle, after the best ideal, a devout and consistent Christian.

§ On the 16th of May, 1648, the Rev. John Campanius returned to Sweden. He had been chaplain of New Sweden since the year 1642, and was a man of much earnestness and application. In addition to his duties as chaplain, Campanius kept a copious journal of his voyage to America and his observations in New Sweden. The Indians frequented the house of Campanius, who never wearied in discussing with them the tenets of his Church, and recorded that he found them able to comprehend the doctrines of his creed. Struck with the patience, aptness, and docility of his pupils, Campanius studied their language, and translated the Lutheran Catechism into the Lenni Lenape dialect of the Algonquin tongue. This book was printed by royal command at Stockholm, in Indian and Swedish, in 1696, in one volume, 160 pages, 12mo; to the text a vocabulary is added, with examples, dialogues, etc.

There is a copy of this Swedish-Indian Lutheran Catechism in the possession of the American Philosophical Society of Philadelphia, and one was owned by Peter S. DuPonceau, LL.D., of the Pennsylvania Historical Society. Clay, in his Swedish Annals, suggests "that the Swedes may claim the honor of having been the first missionaries among the Indians, at least in Pennsylvania; and that, perhaps, the very first work translated into the Indian language in America was the translation of Luther's Catechism, by Campanius."

Presumably, the author of the "Annals" refers to Protestant missionaries; for, not to mention the Spanish priests who came over even with Columbus, and soon made converts, the French Catholics at "Port Royal" (Annapolis, N. S.) began teaching the Micmacs and Abenekis as early as 1605; and the Jesuits were there at public expense as missionaries to the Indians in 1611. De Saussaye founded the mission of St. Sauveur, on the Penobscot, in 1613, which, in August of the same year, Argall, of Virginia, piratically destroyed. There was a mission to the Hurons by Brébeuf, Daniel, and Lallemand, the Jesuit "Fathers," in 1634, and an amply endowed Indians' hospital at Quebec, in 1635. An Indian seminary was founded at Quebec, with money and teachers, the same year, and about the time the Swedes came to "Zuydt River" an Ursuline convent school for Indians was established there."

Five years before Eliot preached to a tribe six miles from Boston, Charles Raymbault and Isaac Jogues, under Jesuit direction, penetrated in 1641 to the outlet of Lake Superior, and preached "Christ and him crucified" to a congregation of two thousand wild aborigines. "Not a cape was turned, nor a river entered," says Bancroft, "but a Jesuit led the way." Dissent is free, thank God! Even dissent from dissent, at last; but history must crown with a just award those to whom, whatever the dogma, THE CROSS meant obedience, patience, and self-denial,—who bore the symbol of a divine humanity to savage men, and, in the speechless death-agony of Indian tortures, offered their cruel executioners the sign of universal love, mercy, and forgiveness!

Campanius and Eliot began labor in the same field at about the same time, and though the work of Eliot was the greatest and most successful, the purpose was identical, and the honor due each is of the same nature. The Swedish chaplain acquired the "Renni Rennappi" language during the six years of his stay on the Delaware, but his translation of the Lutheran Catechism was not put to press until 1696. Eliot began to preach in Indian October 28th, 1646, the Mohegan New Testament was printed in Boston in 1661, and the whole Bible two years later—fifteen years after Eliot began the translation.

The printer's work on this Mohegan Bible—the first Bible published in America—was slowly done by an Englishman, and John Printer, an Indian youth. The work included a catechism, and the Psalms of David in Indian verse. Fifteen hundred copies were printed, at a cost of two thousand dollars; several of them, richly bound, being presented to King Charles of England. "Eliot's Bible" may be seen in the Philadelphia Library, in the library of Harvard College, and a few other like places: few as these copies are, those who can read them are fewer still.

To give an example of the difficulties encountered by Campanius in his translation, it is said that, as the Indians used no bread, he was compelled to translate the Lord's Prayer: "Give us this day a plentiful supply of venison and corn." Eliot, in translating the Biblical account in which the mother of Sisera is described as looking through the "lattice," described a lattice to his Indian assistants, upon whom he was compelled to depend for a word: what must have been his chagrin to find, afterwards, that he had made "the mother of Sisera look out of the window through a wicker-basket trap for eels!" A thorough scholar like Eliot was needed to deal with the synthetical difficulties of a language in which, as no unconverted Indian knelt, the phraes "kneeling down unto him" is of necessity translated and printed Wutappessttukgussunnoohwehtunkguok; yet Eliot translated several works into Mohegan, notably a Mohegan grammar, and an "Indian Logick Primer."

It required the labor of years, says Loskiel, the Moravian missionary, to make the Delaware dialect capable of expressing abstract truth. A new language had to be forged out of existing terms, by circumlocutions and combinations. "Eliot caught the analogies of nature to convey moral truth in his Indian Bible." Each Indian tongue and dialect was a perfectly organized language, expressive of all material things, but there were few words to express aught else; no terms for continence, justice, gratitude, or holiness. It was impossible to translate the doxology into the purely synthetic, absolutely definite Indian tongue, and hence the Onondagas were taught to sing: "Glory be to our Father, and to His Son, and to Their Holy Ghost." Cotton Mather, who based his orthodoxy on witchcraft, gravely stated that he tested the demons around him, who made a pretense of being linguists, with the Indian tongue. These imps, Mather says, frequented his premises, and could well manage Latin, Hebrew, and Greek with ease, but at the Mohegan dialect they shrank back in dismay. The pleasing inference is that the Indians were a people unknown in hell; but the cruel old witch-hunter did not tell the story as a compliment to the Mohegans, but honestly as a fact,—one worthy the most fortunate spiritualist.

Both the Mohegans and Delawares were appreciative of the work done for them by their apostles and catechists. Eliot had three thousand six hundred praying Indians, whom he led like a flock, until King Philip's Indian war, when the men of Massachusetts, mad with terror and despair, turned upon even the inoffensive, praying Indians, broke up their unarmed civilized towns, and drove their innocent red fellow-Christians through suffering to foreign slavery. So perished the hope of John Eliot. The Swedish missionaries sent out by the King and Church of Sweden to the Delaware in 1696 wrote back: "The Indians and we are as one people. They are also very fond of learning the catechism, which has been printed in their language. They like to have it read to them, and they have engaged Mr. Charles Springer to teach their children to read it." And these same people protested alike to Swedes, Dutch, and English everywhere against the sale of rum to their young men.

Few, if any, of the Indians of the Delaware became Christians in the time of Campanius, but afterwards, when broken as a tribe by contact with the whites, the Moravians became the kindly guardians of a part of their people, and many of them joined that church, and settled peacefully and prosperously at "Conestoga," only to be driven from their last home in Pennsylvania by the murderous "Paxton boys," who, coveting their land, killed many of them in 1762. Under the able leadership of their chief, the educated, pious Isaac Still, the remnant of the Delawares emigrated to the valley of the Wabash, "far away" then, as they desired to be, "from war and rum." The last party of about forty started in the fall of 1775. The great tribe had left the banks of the Poutaxit forever. In 1803 Hanna Hannah, last of the Lenapees in the east, died in Chester County. So passed away the peaceful, wise, and influential "original people." "In their dealings with the white man," says Colonel Wm. B. Sipes, in his sketch of the Pennsylvania Railroad, "they were scrupulously honest, and many of them became strongly attached to the early settlers. The treaties they made, which cost them so much and profited them so little, were never broken, and when they had dwindled away, before the advancing tide of civilization, to a mere remnant of a mighty race, they left the burial places of their fathers in search of new homes without a stain upon their honor."

Regretting that the limits of his work prohibit more extended recognition of the faithful Lenni Lenape, the author has chosen a word from their language to grace his title-page, "Scheyichbi" having been the ancient Indian name of New Jersey.

§ The Swedes' Governor, John Printz, writes Governor Winthrop, of New England, in his history, "was a man very furious and passionate, cursing and swearing, and also reviling the English of New Haven as runnigates." The Swedish policy brought Printz into a series of quarrels with the Dutch of Fort Nassau, and they found no exemption from his bad manners. For all that, the Governor of the Swedes was an able man, and not only managed well in the fur trade, but so overslaughed and undermined the power of the Dutch, that in 1649, about ten years after the settlement by Minuit at Paghacking, the Swedes were supreme on the Delaware.

On the 11th of May, 1647, Peter Stuyvesant succeeded Wilhelmus Kieft as Director-General of New Netherland. For several years affairs at Manhattan restrained and preoccupied him, but in 1651 decided measures were taken to reassert the claims of the Dutch on South River, where Stuyvesant proceeded in person. After unsatisfactory negotiations with Printz, the Dutch bought of certain Indians lands five miles below Fort Christina, and at Newcastle, Delaware, they built a fort which they called Kasimir, Fort Nassau being demolished.

Failing to receive the reinforcements he demanded, Printz returned to Sweden, November 7th, 1653, leaving John Papegoia in charge of the colony. But Sweden had not forgotten her colony, but entrusted it to a "General College of Commerce," and in 1653 John Rising, Governor of New Sweden, in command of a strong military force, entered the Delaware, where there had been for some time less than a score of Swedish soldiers. Rising managed to gain possession of Fort Casimir without fighting, and at once fully restablished the power of Sweden, and soon concluded a just peace with the Indians.

When Peter Stuyvesant learned of the "dishonorable surrender of the fort" made by Gerrit Bikker, and of his officers' desertion to the Swedes, with a third of his men, his rage was mighty, and he at once reported the affair to Amsterdam, where his anger was equaled by that of the directors. A Swedish ship, the Golden Shark, entering Manhattan Bay soon after by mistake, was detained "until a reciprocal restitution shall have been made." Meantime, however, Rising wrote home an account of his success, saying that whereas he found but seventy persons in New Sweden, there were then three hundred and sixty-eight who acknowledged his authority, "including Hollanders and others."

On Sunday, September 5th, 1655, "after the sermon," Peter Stuyvesant, with seven powerful vessels and about seven hundred men, sailed from Manhattan, under orders from Amsterdam, for the subjugation of New Sweden. The next (Monday) afternoon the fleet was off Helsingborg, then in ruins; on the 10th of the month the Dutch forces landed near Casimir, which, being much overpowered, surrendered without defense. Rising shut himself up in Fort Christina, and, though closely invested from the 15th, held out until the 25th of September. The Swedish town having been sacked, New Sweden ravaged, and Christina invested by an overwhelming force, Rising, to avoid an exterminating bombardment, surrendered, and the flag of Sweden, which in defense of freedom had waved victoriously in Europe, sank to rise no more in America.

The Dutch forces were recalled to Manhattan in haste to repel an Indian invasion. The conquerers had been in New Sweden three weeks—a body of men twice the number of the entire Swedish population living on the country. Consequently, on the 18th of December, 1655, when John Paul Jacquet arrived at Zuydt Riviere as Vice-Director for Stuyvesant, out of an original population of nearly four hundred but a dozen families remained, and, besides, Fort Casimir was no better than a ruin. On July 12th, 1656, the Dutch West India Company conceded the land from Boomtjes Heuken to Cape Henlopen to Amsterdam, for seven hundred thousand guilders ($266,000); this territory became a colony of that municipality of Holland, under the name of Nieuwer Amstel, the capital being at the present Newcastle. New Amstel was ruled with much rigor; to desert the colony was punishable with death, yet the numerous emigrants sent out by the city could not be retained. A trading post and small garrison were kept up at the Horekill, where in 1662 an Anabaptist "Mennonist" community of twenty-five families settled under the leadership of Peter Cornelis Plockhoy. The Mennonists were a liberal, catholic, tolerant people, and their co-operative institutions were very free and democratic. For several years, owing to disagreements between the authorities of Manhattan and New Amstel, and between both of them and the Governors of Maryland, confusion and distress continued west of the Delaware, and the dissatisfied people were scattered abroad by repeated alarms and panics.

§ The glorious rise and progress of the Batavian Republic astonished the world; the commercial and manufacturing greatness of Holland aroused the bitter and ignoble jealousy of the English. In 1664, in a time of peace and progress, England made a treacherous attack upon the Dutch. On the 8th of September, Manhattan and New Netherland were peaceably but unavoidably surrendered to a piratical expedition which Charles II. of England sent out to place his brother James, Duke of York, in possession of the province of the Hollanders. Sir Robert Carr was sent to take possession of the Delaware. Some defense was made at New Amstel by Hinoyssa the Governor; the place was captured, however, the Dutch soldiers sold into Virginian slavery, and the people plundered, even of their farms in some cases. A boat was sent to the Horekill and the colony there was robbed; among the goods carried off was "what belonged to the Quaking Society of Plockhoy, to a very naile." The court of England tried in vain to justify these acts before the world; they merited the scorn of mankind. Nine years after, even Charles II. repented of his buccaneering; then Holland opened her dikes, and aided by the flood defeated two hundred thousand French troops with twenty thousand man; infinitely bold against desperate odds, the Dutch, at the same time, day after day outfought the fighting ships of Britain, until the shattered fleet, sailing as from an infernal scourge, hid behind the strongest forts, while the revengeful guns of De Ruyter and Tromp bellowed in insolent triumph along the shores of England.

By the overthrow of the power of the Dutch West India Company and the States of Holland in North America, James, Duke of York, became Governor of New Netherland. Before the sailing of the expedition for the conquest of Manhattan, James appointed Nicolls, its commander, his deputy, to act as such after the subjugation of the Dutch colony. Nicolls had been gone from England but a month when, on the twenty-third of June, the Duke of York, well knowing the success of the enterprise was assured by the treachery which conceived it, sold to Lord John Berkeley, Privy Councillor and Baron of Stratton, and Sir George Carteret, of Sattrum, Devon County, Knight, a native of the Isle of Jersey, all the territory now included in the State of New Jersey, which then received the name of "New Jersey," or Nova Cæsaria. James was one of the worst bigots of the English line of kings; all his good qualities, as a man, a prince, a king, were foiled with glaring defects, yet in his honor the name of Manhattan was changed by Nicolls to "New York," the west of the Hudson was called "Albania," and Long Island received the appellation of "Yorkshire;" thus all the various titles of the Duke were foisted upon the country at once—the force of flattery could no farther go.

ATLANTIC BEACH. WHILLDIN COTTAGE. SEA GROVE HOUSE.

The flag of Britain now covered the coast of the Colonies which became the thirteen original United States; freedom and progress were served by injustice in the end, but the people of New York, who imagined the privileges of Englishmen were to be added to the secure possession of their property, soon had reason to sigh for the honest despotism of Stuyvesant, to save them from the extortions of their new and rapacious governors; while the Duke of York and his agents were presently forced to realize in disappointment that the profitable despotism they had planned was impossible among such a people as those they fancied they had made their victims and servants.

By his sale to Berkeley and Carteret, the Duke of York prefigured the outlines of the State of New Jersey, and unwittingly forecast the destiny of a free Commonwealth. The change of government which had made Colonel Nicolls Governor of New York and "Albania" and, as President of the Royal Commission, presumptive potentate of New England, was of vital importance to the people of the Colonies, especially those near New York; and the new administration, appreciative of its opportunities, was not slow to energetically assert its powers.

The citizens of New Haven, who had paid six hundred pounds for lands on the Delaware, and perhaps lost as much more in fruitless expeditions thither, who had remonstrated with Kieft, quarreled with Stuyvesant, and sought the aid of Cromwell, through their General Court, by letter, detailed their grievances to the Royal Commissioners; but the new Governor was too busy to pause to nicely adjust the scales of justice. Ignoring the investitures of the past and the equities of the present, heedless of its own engagements, the government of New York devoted itself to the illegal profit of its officials and the assiduous and flattering service of its ducal patron.

Governor Nicolls, in ignorance of the sale to Berkeley and Carteret, made more than two months before the capture of New Netherland, named New Jersey and the western bank of the Hudson Albania, in compliment to the Scottish title of the Duke of York. This territory he was exceedingly anxious to populate. Tracts of land on Hackensack Neck and elsewhere were granted to parties from New England, who, as required by Nicolls, satisfied the claims of the Indian residents. The Dutch, in 1663, had given a party of Puritans liberty to settle in "Nova Belgia" (New Jersey), with an almost independent charter for a local government, and the settlements under Nicolls were largely the outworking of similar plans by other "Yankee" associations.

The pioneers from New Haven, and those who soon followed them from the east, brought to their new homes the same dogmatic temper and theocratic ideas which characterized the ecclesiastical tyrannies of early New England; but with them they brought also the inflexible resolution and unceasing industry for which the people of that section have ever been distinguished. The New England emigrants soon acquired the influence in New Jersey their pertinacious habits guaranteed from the first, and if the constitution and laws of the new Commonwealth were more favorable to liberty than the primitive enactments of Massachusetts and Connecticut, it was not the fault of the conscientiously stubborn Puritans!

While Nicolls by every means encouraged the settlement of Albania, and noted with pride the multiplying farms and increasing villages from Bergen to Sandy Hook, news came that the action of the dull James of York had disparted his Colony, and conceded the fairest and most promising portion to overreaching speculators. In August, 1665, Philip Carteret entered New Jersey, and by virtue of the provision which, in English law, vested the Proprietary of a colony with jurisdiction, assumed the office of Governor, under the warrant of his father and Lord Berkeley.

Governor Nicolls was much vexed at the unexpected turn thus given affairs, and tried, but in vain, to induce the Duke of York to compel the reconveyance of the territories he had parted with in ignorance of their value. Berkeley and Carteret remained in possession and control, but it was a long time before the duke or his agents, who assumed to hold by feudal tenure, ceased to claim rightful jurisdiction, customs, rights, and paramount sovereignty under the King.

The few settlers Philip Carteret found in his colony were well disposed to receive him as their Chief Magistrate, and when a subsequent Governor of New York invaded New Jersey to intimidate them by a display of the Royal Patent, the sturdy Puritans, without question of the validity of the document presented, referred to Magna Charta as "the only rule, privilege, and joint safety of every free-born Englishman," and stood like a wall for the independence of New Jersey. The beginning of the Commonwealth was but small. On a tract of land once sold by the Indians to the Dutch, and afterwards to the Puritans, four houses stood in the same neighborhood; in honor of Lady Carteret and her kindness, this locality was called Elizabethtown, and in May, 1668, became the scene of a Colonial Legislature and the capital of the Province.

The property of Berkeley and Carteret was almost a wilderness; to induce emigration its owners had sent successful messengers to New Haven to invite the rigid Calvinists to a home on their shores, while, at the same time, the most liberal concessions to liberty were promised whoever should join them in their invasion of the primeval woodlands.

The Governor, the Council, and popular Representatives were to create the laws, persons and property were to be secure, no taxes were to be levied but by the Colonial Assembly, both Proprietaries and people were to unite in maintenance of their mutual rights, even against royal imposition; and last and greatest of all, "freedom of judgment, conscience, and worship" were guaranteed every peaceable person. The power of veto, judicial appointments, and the executive authority were all which was reserved for the Proprietaries. The lands of the new State were to be held under a quit-rent of a half-penny an acre, the payment of which was deferred for five years, or until 1670; and to please the Royal Duke, who was President of the African Company, a bounty of seventy-five acres of land was offered for the importation of every able-bodied negro slave.

As the Dutch patroons had done, settlers were required to base the title to their lands in equity, by a fair and satisfactory purchase of their estates from the Indians.

The compact of New Jersey being ratified by the people, and peace prevailing under the mild sway of Philip Carteret, the province prospered and increased, encouraged by a temperate and salubrious climate, united with a fruitful soil easy of tillage; but in 1670 the quit-rents became due, and then the Puritans, who, in New Haven, had Arthur Smith brought into Court in 1659, and fined fifty pounds, because he expressed some of the "divvilish oppinions" of the "cursed hereticks" the Quakers, developed a peculiar heresy of their own. Referring to their well-thumbed Bibles, from which they were apt to wrench a text to cover any purpose, they argued that Noah was the original proprietor of New Jersey, having in himself and heirs become invested with the same by his landing on Mount Ararat, directly after his protracted voyage in the ark. The title having thus been in Noah, as they argued, followed his descendants. The Indians were lineal offspring of Noah, they bought their lands of the Indians, and hence, particularly as Governor Nicolls had approved the deed and Carteret himself assented thereto, they refused rent which was merely due by the laws of England and their own voluntary contract and agreement.

To save a few shillings, the Puritan farmers precipitated anarchy, drove Philip Carteret from his Governor's chair, and hunted William Pardon, who withheld the records from them, out of the country as if a malefactor. A new Governor was chosen by an irregular assembly of delegates, in the person of James Carteret, a trifling young man, an illegitimate son of Sir George; and while the legal Governor, leaving John Berry as his deputy, voyaged to England for fresh instructions and renewed authority, the revolutionists cultivated their farms in peace, kept the quit-rents in their pockets, and doubtless regarded Noah as a man who had left something very handsome to his family. Great principles dawn slowly on the minds of men, and rightful independence and freedom are evolved, age after age, through the crimes of those who grope toward truth in selfishness and disorder.

While toleration was established in New Jersey and the exercise of freedom urged to the license of revolution, Liberty was exiled from New York, and justice banished that corruption might prostitute the offices of government. There was no popular representation, the Governor and his Council made the laws, decided causes, and assumed executive supremacy; moreover, the functions of government were made means of extortion, and the people were plundered in the name of law and security. Contrary to the stipulations of the surrender, "Even the Dutch patents for land were held to require renewal, and Nicolls gathered a harvest of fees from exacting new title-deeds." That which had been New Sweden was retained under the government of New York, and shared the evils of an extortionate oppression. Governor Lovelace, who succeeded Nicolls in 1667, added to the trials of the people; even the Swedes and Finns became turbulent. "The method for keeping the people in order is severity," said Lovelace, "and laying such taxes as may give them liberty for no thought but how to discharge them." Regardless of the liberties of New Jersey, arbitrary customs were collected at the mouth of the Delaware by the agents of the Duke of York. The people of Maryland invaded Lewestown with an armed force in 1672, to establish the domain of Lord Baltimore on the shores of the Delaware; the country was at once reclaimed by Sir Robert Carr, deputy of Governor Lovelace, as belonging to the Duke of York by conquest.

While all these things took place, the claims of Berkeley and Carteret were reaffirmed in England, and it seemed that trouble was impending for the New Jersey anti-renters; suddenly the political kaleidoscope was shifted by an unexpected hand—Evertsen of Zeeland, commanding a Dutch fleet, appeared in New York harbor, the 30th of July, 1673; again without a blow Manhattan was surrendered, the flag of Holland waved once more over New Netherland. The unjust war upon Holland became unpopular in England, and Parliament refused supplies for its prosecution; peace was declared on the 9th of February, 1674, and the rights of neutral flags were established by the treaty which followed, Holland under the teaching of Grotius having been the first to claim the enfranchisement of the ocean, the freedom of the seas. By treaty, too, England regained the port of New York, with the geographical unity of her Colonies, and the flag of Holland, radiant with victory and honor, was finally withdrawn from the shores of North America.

Under Edmund Andros, the power of James of York was reinstated at Manhattan, October 31, 1674. The narrow-minded duke had learned nothing from experience, and though Andros was a better man than Nicolls or Carr, yet the despotic system which oppressed the people of New York and clutched at the Charter of Connecticut remained. Philip Carteret reappeared in New Jersey, and renewed after a time his argumentative warfare for rights and dues according to feudal law and kingly pleasure, with a people who claimed to hold their lands from Noah, their privileges from Magna Charta, and their faith from private judgment of the infallible word of God.

The Proprietaries of New Jersey sought above all things for profit from their province. Their liberal concession of popular rights was dictated by a policy which, however laudable in its means, looked to the same end gained by the piracies of Carr and the maladministration and extortion of Nicolls. Lord Berkeley was already an old man; as no profit had been derived from his New Jersey property, and trouble was still apprehended from contumacious subjects and disputatious tenants, he became willing to withdraw from the barren adventure.

Where avarice falters in discouragement, and ambition halts in despair, the love of liberty populates the wilderness, and religious enthusiasm builds the institutions of the State. From the time when Charles I. laid his head upon the block in front of his own banquet hall in 1649, the sufferings of "the peculiar people," the Quakers, had been indescribable and universal: whoever was tolerated they were disallowed; they were contemned, insulted, fined, scourged, imprisoned, enslaved, maimed, branded, and hung, even in the New World. In England all classes united to persecute; even the Presbyterians declared that "hell had broken loose" in the person of George Fox, and the mild apostle was forced to denounce them as "exceeding rude and develish." "They were as poor sheep appointed to the slaughter, and as a people killed all day long." And yet, aside from the irregularities of a few fanatics, such as are found in all sects, the offense of the Quaker was only in his spirituality and his democracy. But in the days of Fox and Penn, these were counted worthy of stripes, bonds, and death, by those who worshiped Churches and Kings more than God; and even those who contended to the uttermost for purity of soul, and the right of private judgment themselves, turned like wolves upon a people who gave to the Puritans' version of the rights of man a still more radical translation.

Resolute to bear witness in testimony of the truth of the INWARD LIGHT, ready at all times to be offered up a sacrifice, the Quaker preserved the serenity of his reason, whether he stood amid courts in the presence of kings as a Counselor and Friend, or perished from hunger, cold, and neglect amid the frozen filth of dungeons. He who "affirmed" himself the peer of peers, wore his hat as only a peer by law might do; the "Friend" was ready with his "thee" and "thou," and other titles he would have none; but "plain speech" was not impertinent language, and formal dress meant other things than eccentricities of character.

Determined on freedom, the Friend was not bent on useless martyrdom, and while Fox journeyed as a missionary, and Penn traveled as a preacher, the iconoclasts cast about for an asylum for the persecuted, a land where "Salem" might be founded, where "the Holy Experiment" might be tried, and, God willing, "Philadelphia" arise to welcome to the "city of brotherly love" the universal tribe of man. Penn traversed Europe, Fox the colonies of America; nowhere was there to be found rest and peace, except perhaps in the narrow confines of

Rhode Island. New Netherland turned aside from the policy of Faderlandt, and half-tolerating Lutherans; Stuyvesant had only imprisonment, labor in chains, and the dungeons of Fort Amsterdam for "the new, unheard-of, abominable heresy, called Quakers."

At last light dawned from afar, and in New Jersey there was hope. Edward Byllinge, by John Fenwick, as trustee for himself and his assigns, bought of Lord John Berkeley, in 1675, for a thousand pounds, one undivided half of New Jersey; under this indirect purchase misunderstandings arose, but they were managed by the arbitration of William Penn, to whom, with Gawen Laurie and Nicholas Lucas, Byllinge finally assigned his property for the benefit of his creditors.

In June, 1675, Major John Fenwick, claiming his own right as an associate in the purchase with Byllinge, arrived in the Delaware in the ship Griffith, "with a large company and several families." The arbitrators had assigned one-tenth of Byllinge's purchase to Fenwick, with a sum of money as his share; he assumed the character and style of Lord Chief Proprietor. Near where the people of New Haven had settled on Varcken's Kill, not far from the site of the Swedish fort Helsingborg, and where the relics of unknown pioneers were found, the colony of "plain John Fenwick" also selected their location, and, feeling secure at last, landed upon the peaceful shores and bestowed the name of "Salem" upon the place. Byllinge had failed, and, in the interest of his creditors, the nine-tenths of one undivided half of New Jersey, left to his estate, was offered for sale in decimal shares of tenths and hundredths; to carry out the purpose of an asylum for the persecuted, these shares were largely taken up by Quakers.

To found his colony, John Fenwick had borrowed money of John Eldridge and Edmund Warner, giving his tenth of the Byllinge purchase as security, with the right to sell lands therefrom to their satisfaction. Eldridge and Warner conveyed their claim to the trustees, Laurie, Penn, and Lucas. Fenwick still asserted himself in all the qualities of Lord Chief Proprietor, refusing to abide by the results of arbitration. The rights and claims of Fenwick were a sore trial to Penn, and he and his associates have been accused of duplicity in regard to the matter, how justly or unjustly still seems a matter of dispute; however it may have been, Fenwick abode in his place, and as long as he lived gave token of an uncompromising and dauntless, even if, at times, impolitic and arbitrary spirit.

As soon as the matter of ownership was adjusted, the Quakers secured from Carteret a division of the estate. Anxious to come into possession of their territory, where they could institute a government, the Friends haggled not for advantage, and Carteret, conscious of having the best of the bargain, readily fell in with their proposals. The line of division ran from Egg Harbor to a point on the Delaware River, under the forty-first degree of north latitude, and near Burlington; the lands to the north and east were to be left to Carteret, and those to the southward and west, under the name of WEST NEW JERSEY, became the property of Quaker associates.

Long accustomed to endure suffering, competent as critics and preachers, "the peculiar people" were now to be more severely tested; they were required to build, to organize, to govern and enjoy. Consulting among themselves in England, the Friends evolved their scheme of government. "The CONCESSIONS are such as Friends approve of," wrote the Quaker Proprietaries to those already in their land of rest. "We lay a foundation for after-ages to understand their liberty as Christians and as men, that they may not be brought into bondage, but by their own consent, for we put THE POWER IN THE PEOPLE."

The basis of the Quaker State was democratic equality; methodically and clearly the "agreements" stated the sublime affirmations of the Quaker, and in harmony therewith promulgated the "fundamentals" of the highest form of actual government the world has ever known. Freedom of conscience, the ballot-box, equality before the law, the right of assembly, freedom of election, freedom of speech, freedom of the press, popular sovereignty, trial by jury, open courts, free legislatures, all these were provided for in West Jersey, in March, 1677. What more? No poor man could be imprisoned for debt, none held as slaves; there was free access to the courts, where each man might plead for himself; the judge, an appointee of the assembly for two years only, merely announced the law, the jury gave both the verdict and the sentence; where Indians were concerned the natives were to make half the jurymen. The statutes prescribed were admirable and consonant with the Constitution, the whole wise, just, and discriminating, full of justice, benevolence, and protection even to the humblest denizen of the aboriginal woods. The helpless orphan became the ward of the State, and the child of misfortune was educated at the cost of the Commonwealth.

The honor and fame of William Penn are borne toward future ages with the progress of the mighty State that bears his name; but, let it be remembered, in West Jersey his inspired mind and benevolent heart first wrought out his model of a state, and there, and there alone, his will and his purpose became the law and rule of a happy people. Every acre of New Jersey has been fairly bought of the Indian tribes. West Jersey is unstained by Indian blood. "You are our brothers," said the sachems; "we will live like brothers with you. The path shall be plain; there shall not be in it a stump to hurt the feet." "Their ways were ways of pleasantness, and all their paths were peace."

The "holy experiment" had been established, and thus far was successful; troubles and trials came at length, but new precedents met novel emergencies, and staid historians who describe the time break forth in poetic rhapsody to tell of the happiness of the people. "The people rejoiced under the reign of God." "Everything went well in West New Jersey."

Meantime, the trustees of Sir George Carteret grew tired of Colonial burdens and trials without return, and proposed the sale of East New Jersey. The estate was purchased by William Penn and eleven others, the first and second days of February, 1682, for three thousand four hundred pounds; possession was taken in November, 1682, by Deputy Governor Thomas Rudyard, for the Association. New Jersey was now entirely in the possession of Friends, but in East Jersey were found a large number of "sober professing people" of the Calvinist persuasion, and sound policy seemed to require a more varied board of proprietaries. Accordingly, each Friend selected a partner, and, to the twenty-four, a new patent was issued by the Duke of York. The King also confirmed all the transactions by declaration in November, 1683. The partners were not all Quakers, but one of them, who was a Friend, the able Robert Barclay, of Urie, Scotland, was made Governor, and afterwards became Governor for life.

While important events thus followed each other in New Jersey, William Penn secured his grant of Pennsylvania, and, late in the autumn of 1682, he held a meeting at Shackamaxon to which the Indians of Pennsylvania were invited, and where the spirit moved Penn to preach a Quaker sermon,—the same gospel George Fox announced to Cromwell, and which Mary Fisher delivered among the armies of the Turks and bore to the Sultan, "Commander of the Faithful." "We are all one flesh and blood," said Penn. "We will live in love with William Penn and his children as long as the moon and the sun shall endure," answered the "savages;" and *they* kept their word, and long treasured the tradition of that day's speech from *Onias*, the great Father of the *Quekels*. Says Bancroft, "Not a drop of Quaker blood was ever shed by an Indian."

But the affairs of Pennsylvania became too vast for personal superintendence, and the agents of Penn in the purchase of lands, in making of treaties, often forgot his gospel and disregarded the wishes of his gentle soul. In time "the world's people" rolled in on Pennsylvania like a flood; professing obedience to Biblical law, and denouncing "vengeance on the heathen," they themselves selfishly trampled on all law, human and divine, and, under the hypocrite's cloak of zeal for the glory of God, defied the rights of Penn and his assigns, overrode the laws of the Province, intruded without warrant upon the lands of the tribes, and imbrued their hands in the blood of the Indian with every circumstance of base atrocity, even to those who knelt at the name of Jesus and shared with the Moravian saints the bread and wine of the Christian Sacrament. In 1682, Penn promised the Indians, "No advantage shall be taken on either side, but all shall be openness and love."

In 1685, the agents of Penn shamefully defrauded the tribes of their lands to the Susquehanna, and, in 1764, John and Richard Penn, the sons of "*Father Onias*," sanctioned Lieutenant-Governor Cadwallader's offer of one hundred and fifty dollars for the scalp of an Indian, and one hundred and thirty-four dollars for scalps which bore the hair of a squaw! The Pennsylvania Quakers, many of them, labored faithfully and not in vain in the cause of justice and mercy, as they had light, but the student who seeks the logical issue of the principles of Fox and Penn starts back in grief and horror from the blood-stained soil of Pennsylvania, to follow the record of events east of the Delaware.

Theological predestination means political democracy. Quakerism is the democracy intended, and yet predestination alone separates the Friend and the Calvinist. "The nearer the relation, the worse the quarrel," and in all the weary years, from George Fox, in 1649, to the death of Charles II., in 1685, Presbyterians in England were the persecutors of Friends; and in Massachusetts the Puritans ordered that the ears of the Quakers be cut off, and their tongues bored with a red-hot iron. They were Calvinists who, in Boston, in 1659, put Marmaduke Stephenson, William Robinson, and William Leddra to death on the gallows for preaching Quakerism in Massachusetts, and hung Mary Dyar on Boston Common, the same year, for the same offense!

Cromwell died, the Stuarts were restored, Charles II. reigned for the quarter of a century; the zealous fanaticism of the Calvinist Roundheads was succeeded by the superstition of the divine right of kings, the last deepened by the excesses of the first. Monarchy was absolute in church and state in the last days of Charles II., "Independents" were marked for destruction, and "Presbyterians,"—they who since the time of Edward VI. had originated each struggle for popular freedom, they who always dreamed of republics, whose creed taught insubordination as a dogma,—what had they to expect? It was in Scotland that the policy of the Stuarts bore its ripest fruit; there the crime of Cromwell in the execution of King Charles I. was ten thousand times revenged. Of the Cameronians, of the Covenanters, of the Scottish Presbyterians, what can be said? Nothing exceeded the cruelty, the brutality, the mad, exterminating barbarity visited upon them, except, forever, the fortitude with which they confronted those who slew them! The magistrates of Boston, in 1659, were tender nursing mothers—angels of mercy—compared to Claverhouse and Lauderdale and Jeffreys, the minions of episcopacy and the king.

Atrocity incited insurrection, but the adherents of Monmouth were borne down, and the penalties of treason superadded to the inflictions of persecution. All who had ever communed with rebels were condemned; twenty thousand lives awaited the executioner, safe only in the forbearance of the informers. In the name of law, the common dragoons, the rank and file of the soldiery, were made magistrates and judges over families of rank and wealth and women of culture, as well as the peasantry of the mountains. The discretion of the ruffians themselves furnished the instructions of this banditti, but royal mercy moderated their rigor. Summary murder was forbidden, and women were to be allowed to die without dishonor; no other restriction was imposed. To whom, among the bloodhounds of vengance, should a maiden make her complaint of outrage, and when have the dead returned to convict their assassins?

There was not room in jail for all the Covenanters; the prisoners were sold into plantation slavery, and the price of blood shared by royal favorites. Presbyterians were hunted like vermin, with dogs and guns, by mounted men led on by swarming spies; it was death to house them, death to throw them bread, death to listen to complaints of theirs; did a wife, a husband, a father, a child, a parent, comfort their own kind, death was the doom of both the sufferer and the friend. It was more than human nature could endure, and the bewildered, despairing victims of an infernal crusade turned at bay and threatened retaliation. Such is the courage of the hunted, bleating ewe, when bloody wolves rage round the mangled flock. The threat of resistance was answered by the order for massacre. As they labored, as they prayed, as they journeyed, as they fled, the Covenanters were shot down; their estates were plundered, their houses burned, their families hurried away to distant colonies.

James II. came to the throne; he only added the aggravation of a delusive pretense of clemency to the miseries of the people. The victims of cruelty sought in flight safety from death; every day companies of fugitives were arrested by the troops; juries of soldiers trying them beside the highways, they were condemned in a body and shot in heaps together. Beside the sea women were tied to stakes at ebb of tide, far out upon the strand; the pitiless tide returned by slow degrees, and, mocked by the ribaldry of the troops, who laughed at the amusing spectacle, they were gradually and agonizingly drowned. The dungeons were crowded with men; for food, for water, for air, they prayed in vain; starved, choked with thirst, or suffocated, they died in breathless torture. But the Government of England was not merciless. When the dungeons would hold no more, living or dead; when the assassin tired of murder; when only suspicion indicated a victim; when a whim suggested forbearance, then shipload after shipload, in crowds the wretched, plundered, ill-provided exiles were sold and exported to America. Still monarchy and episcopacy laid their hands upon them as they left their native land; some of the men were allowed to retain a single ear, but others were deprived of both, while upon the cheeks of fair women and matrons the branding-iron was often deeply set, while a royal mandate crossed the Atlantic to forbid mercy or mitigation of their slavery.

Now how might the Quaker exult in his happy home between the Delaware and the sea, and, secure in the immunities of his own freedom, reflect that the Lord had revenged his wrongs upon those who had joined with the multitude to do him evil! Had the FRIEND been other than "friendly," now was the time to satiate his malice, for the groans of his tormentors were in his ears, his eyes witnessed the full measure of their suffering.

But what revenge may men take to whom the INWARD LIGHT dictates a rule of action? During the reign of Charles II., James, then Duke of York, was the friend of Admiral Penn, and, just before the admiral's death, pledged him the same regard for William Penn, his son. When the duke came to the throne as James II., William Penn had great influence. The king, a bigoted Roman Catholic himself, stood in need of toleration in England, where the Established Church, though persecuting the Covenanters to the death, hated Romanism more. The Papist king persecuted Protestant dissenters to win the political favor of the Church of England. The plea of Penn was for toleration, not for himself alone, but for all; he averted persecution from Roman Catholics on the one hand, and restrained as far as in him lay the storm of rage which overwhelmed the Presbyterians. He was accused of Jesuitism, popery, and treason in consequence, and, though disproving every charge, became suspected by men of all parties because he was active in defense of the common rights of each.

When Penn moved in the purchase of New Jersey, it was not merely as an asylum for Friends, but to provide a home for all who suffered for conscience. No sooner was New Jersey under Quaker control than a fair and reasonable description of it was published, and an account of its free and tolerant institutions forwarded therewith to Scotland. The Quaker founded a State in freedom, and made it the home and asylum of those who had deprived him of liberty and life. And this was the revenge of the men with broad-brimmed hats, who "theed and thoud" alike the plowboy and the monarch. To be true to principle regardless of persons, to resist not evil, but return good for evil,—such has been the teaching of the INWARD LIGHT. In Judea or New Jersey the gospel was the same: "Do good unto them who despitefully use you." Well had the Quaker heeded the teacher, and well had he comprehended the lesson!

Convinced of the purpose of the Government of England "to suppress Presbyterian principles altogether," and perceiving that "the whole force of the law of this kingdom is (was) leveled at the effectual bearing them down," the ruined Scotch Presbyterians, in whose souls a sense of duty to God forbade conformity to human assumptions, were ready, as soon as the way opened, to abandon even "bonnie Scotland," since apostasy alone could ransom their lives in their native land. A number of Scottish Covenanters arrived in East Jersey in 1682.

George Scot, of Pitlochie, was a leader among the emigrants. "A retreat, where by law a toleration is allowed," said he to his neighbors and fellow-suffering countrymen, "doth at present offer itself in America, and is no where else to be found in his Majesty's dominions." To America, to East New Jersey, came George Scot and family, and about two hundred others, in 1685. During the following year, after the Duke of Argyle had been put to death under mere pretense of law, Lord Neill Campbell, the brother of the murdered nobleman, became, by purchase from Sir George Mackenzie, one of the Proprietors of East New Jersey. Lord Campbell sent over a large number of settlers, and, coming himself for a time, acted for some months as the Chief Magistrate. Lord Campbell was succeeded in office by Alexander Hamilton; the power of the Proprietaries was inconsiderable. Monarchy had no call to the New World, there it existed only by its feudal shadaw; feudalism was already outworn in Europe, and of the outworn shadow Proprietary Government "was that shadow's shade."

But what need of thrones, of nobles, of titles, of cumbrous institutions to this people? They who held themselves as sons of God, coheirs with Christ; whose glory was foreordained in the eternal councils of the Almighty, and their names written in the "Lamb's Book of Life," from the foundation of the world—the elect, the redeemed, the sanctified, the persevering saints; the children of the COVENANT! Virtue, education, courage, experience, they had them all; religion inspired them, the love of liberty controlled them; nature gave them the harbors of Scotland, the fertility of England, and the climate of France; with the forests, the game, the fish, the fruits, and the freedom of America, beside the "curious clear water" which flowed in abundant brooks and rivulets along the healthful vales of New Jersey. The ocean rolled between them and persecution, between them and every hostile tribe abode peaceful Quakers, who practiced a blessed white magic upon the wildmen, and transformed them to philanthropists. There was a world of room, great flocks of sheep pastured beside the roads of imperial width, and troops of horses fit to mount the squadrons of a king bred and multiplied uncared for in the woods. Not thus grew the many children of the Scottish Calvinists, as in New England free schools were soon provided for, and education and moral training cared for the coming generation.

Indians, Puritans, Quakers, and Covenanters held in peace and universal prosperity the soil of New Jersey. Toleration is a narrow word: they met on the broad platform of equal rights, of judgment, and mutual union for the common weal and wealth. America welcomed every sect, predominant bigotry became impossible. The pioneers of New Jersey were strong souls with varied thoughts; there moderate counsel has prevailed, and seeking to preserve the rights of each, the people have maintained the noblest freedom, and fostered the prosperity and happiness of all.

James II., fickle and inconsistent in everything but personal selfishness and the greed for arbitrary power, had no sooner reached the throne than he undertook to make the colonies "more dependent." In New York the honest advice of Penn, which was demanded in 1682 by the duke, won for that State her "charter of liberties," but James, as king, trampled upon his engagements as duke; tyranny returned in New York, and the Proprietaries of New Jersey were compelled to surrender their rights of jurisdiction. Sovereignty over New Jersey was merged in the crown in 1688. For three years after 1689 East New Jersey had "no government whatever." For twelve years the whole of the province was without settled administration or recognized Governors. The Proprietors, anxious to preserve the forms of law, tried in vain to exercise a power they had renounced, but, divided among themselves, they but divided the people, the courts and the records shared the confusion, politicians pushed their disagreements, but the virtue of the people preserved society.

The crimes of James II. against the Dissenters failed to secure for him, as a Papist monarch, the alliance of the Church of England; in revenge, he proclaimed equal franchises to every sect; toleration was to weaken the episcopacy, and reconcile the English to Rome; it brought William of Orange to the throne of Britain, in 1688, and drove James II. into poverty and exile. The advent of William was a great revolution in England: it secured toleration for all Protestants, and established the rights of the subject on the basis of English law.

When, in 1702, Queen Anne came to reign, matters in New Jersey were still unsettled, the law officers of the crown questioned the selfish arrangements of those who had for gain bought out original Proprietaries, and Parliament threatened interference in a province "where no regular government had ever been established." The Proprietaries, to avoid litigation which might have endangered their ownership of land as well as their pretended rights as Governors, surrendered their claims to jurisdiction, unreservedly, before the Privy Council of England, April 17, 1702. As simple owners of land, the Proprietaries managed to retain their full rights, and became merged in the landholders of the province, their titles descending unimpaired to their assigns and heirs. After the surrender of the Proprietary, the whole of New Jersey was governed by a royal Governor, it never again obtained a charter; power was monopolized by officers under royal instructions, and toleration denied Papists; "no printing-press might be kept," or any publication made without license; meantime, the traffic "in merchantable negroes" was stimulated by every means in the power of the provincial government, under instructions from the throne. Thus the power of monarchy found the refugees in the forest; but Quakers, Puritans, and Presbyterians united in a stubborn, able, and yet orderly struggle for former freedom. Peacefully but sternly the debate had begun, to end in making New Jersey a sovereign state, in an independent confederacy.

The disputes as to jurisdiction, titles, etc., between the Duke of York and the proprietors of West Jersey—the trustees of Edward Byllinge—were decided by Sir William Jones, in 1680, in favor of the proprietors; but the duke, in his new patent, unwarrantably made Byllinge hereditary Governor. The nomination was unprovided for in the constitution of West Jersey, but to avoid further trouble a precedent was made, and Byllinge elected; he, however, continued in London, having little influence in the province.

In 1687, Byllinge died, and Doctor Daniel Coxe, of London, himself a principal West Jersey proprietor, bought the claims of the heirs of the Governor, and undertook to organize a government, by adopting the constitution of England in place of the original Quaker Concessions. Near Town Bank, Cape May, on Coxehall Creek, Dr. Coxe built "Coxe Hall" for a residence; on the draft of a primitive survey, made in 1691, the edifice appears, adorned with a tower or spire, quite in contrast with the original cluster of whalemen's cottages, not far away. The above-mentioned survey was made by John Worlidge and John Budd, who, coming down from Burlington, laid off ninety-five thousand acres of land in Cape May County for Dr. Coxe. The people were not inclined to co-operate with the new Governor in his designs, and he labored in vain to establish the feudalism of England on the shores of the Delaware; still he continued to speculate in Indian lands, and a few of the original settlers of Cape May secured their estates directly from his agents.

Dr. Coxe was a man of vast enterprise and unbounded yet not unreasonable ambition, and was concerned in the attempt to found an English province in Louisiana, which was rendered futile by French pre-occupation. In 1692, the "West Jersey Society," an organization of forty-eight persons combined for the purpose, bought of Dr. Coxe, on the 20th of January, the whole of his claims to lands and jurisdiction, paying therefor the sum of nine thousand pounds sterling. The Society put their newly acquired lands in market, in tracts to suit, at moderate rates, much to the public benefit; as they sold in *fee simple*, independent landlords and small farmers became numerous, and the foundation of a democratic state was laid in a freeholding population.

§ Prominent in geographical position, remarkable in its natural features, and especially fortunate in climate, Cape May attracted the notice of the earliest navigators of the adjacent seas, and was soon celebrated by the explorer and naturalist. In 1641, the site of Sea Grove or its vicinity was referred to as a promontory, and Campanius wrote of dangerous shoals off Cape May, no longer in existence. Whatever improvements natural causes have made in the mouth of the Delaware, the sands have been piled over against Henlopen, and with the shoals have gone the dunes, or beaches, which made the point a promontory.

The historian of Cape May finds no records of white men before 1685; then Caleb Carman was appointed Justice of the Peace, and Jonathan Pyne made Constable by the Assembly of New Jersey, thus indicating a pre-existing population. Cape May was cut off from the north by vast, dense, impassable cedar-swamps, extending from the sea-shore to the bay, and must, in prehistoric days, have been a wild and almost inaccessible place. The earliest known inhabitants of Cape May were, of course, Indians, and, according to Captain Samuel Argall, in 1610 they were numerous. As a fishing station, the cape may have been occupied at any time for the last three hundred years, and the pirate and slave-hunter preceded even the fishermen.

The history of the Delaware valley indicates clearly that the first residents of Cape May were refugees,—persons who, to escape servitude, oppression, or debt, domiciled in the wilderness. The Swedes, who sometimes visited the cape for eggs and to kill geese, solely for their feathers, had in their colony men bound to penal slavery; some of them became fugitives among the Indians. When, in 1642, the New Haven colony on Varcken's Kill was broken up, some of its members remained on the Delaware, and subsequently New England vessels harbored at Cape May, fishing and trading for furs,—an illicit business for them in the judgment of the Dutch and Swedes. One such vessel was robbed and her crew murdered by the Nanticokes, near Swaanendael, in the spring of 1644. She had spent the winter at Cape May, and went over for beaver-skins. From New Sweden, from New Amstel, from the colony at the Horekill, as may be recalled, varied causes at different times scattered the people; most of them fled to Maryland, many crossed into New Jersey, and some, doubtless, reached Cape May.

In his "Early History of Cape May County," Maurice Beesley, M.D., referring to the probabilities of prehistoric settlement, writes: "It would seem probable, inasmuch as many of the old Swedish names as recorded in Campanius, from Rudman, are still to be found in Cumberland and Cape May, that some of the veritable Swedes of Tinicum or Christiana might have strayed or have been driven to our shores. When the Dutch Governor, Stuyvesant, ascended the Delaware in 1654 (5), with his seven ships and seven hundred men, and subjected the Swedes to his dominion, it would be easy to imagine in their mortification and chagrin at a defeat so bloodless and unexpected, that many of them should fly from the arbitrary sway of their rulers, and seek an asylum where they could be free to act for themselves without restraint or coercion from the stubbornness of Mynheer, whose victory, though easily obtained, was permanent, as the provincial power of New Sweden had perished forever."

Pieter Heyser began whaling in Delaware Bay in 1630. When it became a regular business at Town Bank is uncertain; there was a fisherman's colony there from New Haven and Long Island of considerable numbers, and living in houses, before 1691: outstaying the whales, they took up farms, resorted to other pursuits, made themselves homes, and founded some of the best families in New Jersey. The first account of a visit to Cape May was published in a "Description of New Albion," written by Sir Edmund Plowden, under the name of "Beauchamp Plantagenet," which appeared in London in 1648. Plowden reproduced a letter from Lieutenant Robert Evelyn. "Master Evelyn" left England with an expedition for the Delaware in 1634, and probably made his exploration of the cape soon after. Others had observed Cape May,—Hudson in 1609; Argall, 1610; Cornelius Hendricksen, 1616; Dermer, 1619; Cornelius Jacobsen, May (1614?), 1620; Hossett and Heyes, 1630, and De Vries in 1631; besides a party of eight, sent to explore the bay, in 1632, by Governor Harvey, of Virginia, who were killed by Indians.

Cape May County was instituted the 12th of November, 1692. There were five members of Assembly allowed it; the next year a quarterly court, for cases not exceeding twenty pounds, was decreed by the Assembly of New Jersey. The first court was held at "Portsmouth" (Cape May Town, or Town Bank), on the 20th of March, 1693. The Grand Jury having been charged, found "it necessary that a road be laid out, most convenient for the King and county; and," said they, "so far as one county goeth, we are willing to clear a road for travelers to pass," as if the guardians of the county saw, prophetically, how much their district was to owe its future growth and prosperity to the appreciative health- or pleasure-seeking traveler. The tax levied in 1693 was forty pounds sterling, with the considerate proviso that produce should be taken at "money price" in payment. One of the first acts of the court was an order that "no one shall sell liquor without a license," the traffic and use of rum having already, as usual, been the cause of much trouble.

Of the settlers at Cape May in 1685, and of those who came for some fifteen years after, the majority were attracted by the whale fishery in the bay of Delaware. It is shown by reliable records, that whaling was the business of Christopher "Leamyeng" and his son Thomas, of Cæsar Hoskins, Samuel Mathews, Jonathan Osborne, Nathaniel Short, Cornelius Skellinks, Henry Stites, Thomas Hand and his sons John and George, John and Caleb Carman, John Shaw, Thomas Miller, William Stillwell, Humphrey Hewes, William Mason, John Richardson, Ebenezer Swain, Henry Young, and many others. In looking over the colonial records of New Haven, in the first years of its existence, the reader meets most of these family names, and the Long Island whalemen were of the same stock. The same names are found to-day on the books of New England ships; they are people of Newport, of

BIRD'S-EYE VIEW OF SEA GROVE. FROM THE OCEAN. 1876.

Nantucket, of New Bedford, and New London; the world had no such dauntless mariners as the whalemen of New England and Cape May.

The purchase of the rights of Dr. Coxe being made in 1692, the West Jersey Society, as proprietors, to prevent confusion, nominated Andrew Hamilton, the former deputy of Governor Barclay, to be Governor. The people at large acquiesced, and the General Assembly of New Jersey passed an act to cure all defects in law and practice. The law officers of the crown, however, refused their sanction to such legislation, and the lords of trade claimed New Jersey as a royal province. The basis of government continued unsettled, and, in 1702, the New Jersey proprietors surrendering their claim of jurisdiction, as has been noted, continued to hold their lands under the Royal Governor, Edward Hyde, the weak, yet arrogant, "Lord Cornbury."

Much of the difficulty in establishing government in New Jersey arose from the factious opposition of parties who wished to avoid the payment of quitrents, and prevent adverse decisions against their insufficient invalid land titles. Not altogether wrong in equity, perhaps, these persons still evaded the courts, and by their interested captiousness defeated the plans of moderate men, did wrong to their neighbors, and kept the province in a chaotic state, until it lost its charter, and passed under the shadow of arbitrary power. In Cape May County there was little dispute about titles to land; Coxe held most of the soil, though but five sales were made by his agent George Taylor. The West Jersey Society continued the sale of lands for sixty-four years, and by 1756 had disposed of most of their estate. Doctor Johnson, of Perth Amboy, was the principal agent of the Society at the time, and Jacob Spicer (2d), in a negotiation in which the wine-bottle is said to have betrayed Johnson into forgetfulness of his employers' interests, bought the remainder for the insufficient sum of £300; at his death, Johnson, seemingly conscious of his unfaithfulness, left the Society a thousand pounds conscience-money.

By English feudal law the West Jersey Society became, through their purchase from Coxe, invested with a monopoly of the natural privileges of Cape May: none could legally fish or hunt without their consent; the deeds given by the Society did not convey these natural privileges, and much anxiety was felt about the matter in time, although the Society prohibited none from oysters, fish, or game. An organization was created in 1752 to secure the natural privileges for public use, but delay occurring, Spicer forestalled their action by his jolly bargain with Doctor Johnson, and by so doing provoked a quarrel with his neighbors, which was discussed in a public meeting at the Presbyterian meeting-house, March 26th, 1761. The following June, Spicer, who never sought to prevent his neighbors from using "the natural privileges," offered to sell his whole landed estate in the county, excepting his farm at Cold Spring Neck, and the natural privileges, except a right for his family in the same, to the people of the county for £7000, but his offer was declined. "I was willing," wrote he, "to please the people, and at the same time do my posterity justice, and steer clear of reflection."

It must have been an unpleasant affair for Spicer to be at variance with the people whose representative he had been for seventeen years. An active man of exemplary habits and comprehensive mind, Spicer was twenty-one years in the assembly, being first elected in 1744: he was appointed by the legislature one of the commission which met in 1758 at Crosswicks, and then at Easton, to extinguish by special treaty the Indian title to lands in the State. By the work of this convention New Jersey gained the title of "the great doer of justice" from the Delaware tribe of the Lenni Lenape.

Jacob Spicer (2d) dying in 1765, his son Jacob conveyed the natural privileges to a corporation organized by the legislature; thus feudal rights were recognized; besides, an East Jersey court gave a decision in favor of the rights of the proprietors; an appeal was taken, however, to the Supreme Court of the United States, and the verdict below reversed, and the State made the proprietor of the privileges of the water for the use of the whole people. Thus the last trace of feudalism disappeared, and the visitor enjoys the sports of Cape May, thoughtless of the "natural privileges" about which so much unavailing pother was made so long ago.

§ The earliest historical settlement in Cape May County was that of the whalemen at "Town Bank," a bluff the visitor at Sea Grove can see, as part of an unequaled view, from the observatory over the Pavilion. From the tower Town Bank is the highest ground in sight, lying some four miles away due north, and on the shore of the bay. Before 1700, most of the land taken up was in that vicinity. The marine taste and habits of the people coming afterwards are attested by the fact that they settled altogether along the bay or sea, heedless of the quality of the soil.

It is only within the last generation that the inland portions of West Jersey have attracted the attention its resources justify; the unexampled growth of such a town as Vineland, within less than a score of years, is an indication of the results of enterprise in that region; still, the Jersey shore will have its share of residents, especially in summer, for reasons which are palpable to all who observe them from one of the beautiful sail-boats, which the tourist always finds near Sea Grove, "well kept, ataunto, spruce, and gay," awaiting his pleasure.

The waters of Cape May are magnificent for varied sailing. The sounds are as smooth and placid as a garden pool; there the most timid may venture, cruising without a fear, yet the sea breeze sweeps across them, damp with the spray of the adjoining breakers, and the voyage may be extended all the day.

For the many "not afraid in a boat," the bay and roadstead are a safe and free expanse of pleasant waters; while those who love the breeze, the blilow, and the spray, in all their ocean sublimity, have before them the broad Altantic, clear of reef or island for three thousand miles to "the far-off Azores," and beyond for hundreds of leagues to Lisbon, Portugal, and old Spain.

As late as 1706, the only routes from Cape May to Burlington were by the river, and over bridle-paths which led hither and thither across and through the forests, swamps, and marshes. Thomas Chalkley, an English Friend, rode from Cohansey to Cape May, 2nd month, 1726, "through a miry, boggy way, in which we saw no house for about forty miles, except at the ferry;" "that night," says his journal, "we got to Richard Townsend's, at Cape May, where we were kindly received." At Townsend's, at Rebecca Garretson's, at John Page's, at Aaron Leaming's, Chalkley held satisfactory meetings; he stopped two nights with his wife's brother, Jacob Spicer, and journeyed to Egg Harbor. "We swam our horses," wrote he, "over Egg Harbor River, and went over ourselves in canoes." The difficulties of travel may have been one reason why the people of Cape May chose Peter Fretwell, a Quaker resident of Burlington, to represent them in the assembly in 1702, and for twelve years after; it seems a strange proceeding any way, but all New Jersey was full of odd political devices in the early days.

As early as 1698, Richard Harvo, of Cape May, owned a sloop; and in 1705, Captain Jacob Spicer sailed the sloop Adventure of sixteen tons, John and Richard Townsend, owners, as a packet between Cape May, Philadelphia, and Burlington, under a license from Lord Cornbury. In 1706, another sloop, named the Necessity, was built and owned at Cape May by Dennis Lynch, from which time the marine increased until in fifty years there were numerous small vessels trading from Cape May County to Oyster Bay, Long Island, to Rhode Island and Connecticut, and to Philadelphia. The vessels going east generally carried lumber, while oysters and produce of various kinds found a market up the river. Jacob Spicer (2d) owned a vessel he sent to the West Indies, and he shipped much corn taken by him in barter for general merchandise. In 1750, the Delaware pilot-boats were pinked stern boats, sharp at both ends; a usual size was twenty-seven feet keel and eleven feet beam; the "pinkie" was the lineal progeny of the "whale boat," and, when in familiar hands, one of the stanchest craft that ever rode a wave.

It is easy to imagine the slow but yet actual improvement in the means of transportation around and from Cape May; but it was not until ~~1872~~ that change amounting to a revolution took place in the means of travel. In that year Captain Wilmon Whilldin put the first steamboat on the route between Cape May and Philadelphia. Though regarded almost as a miracle, the boat was a modest craft compared to those which glide along the Delaware now; "the longest day in June" was almost too brief for her to make the trip from the Cape to the city, running "between sun and sun." But other boats were soon put on the route, which reduced the time of travel while enlarging the accommodation. There are few finer trips than that down the lovely Delaware; the land disappears at last as the mid-waters of the bay are crossed, and Sea Grove comes in view, often from a deck that reels merrily beneath the feet of the voyager. For those to whom even the bracing air of the bay has no charms, unless inhaled from the shore, there has been provided another line of travel.

The West Jersey Railroad was completed to Cape May in 1866, and since then each year has added to the excellence of the road itself, while the time consumed in the journey has been reduced to the minimum consistent with safety. Cars of the most complete construction and luxurious finish are run, including Woodruff's Silver Palace Drawing-room Coaches, and a degree of care and courtesy is evinced by all engaged in train-service, which render the journey of only two hours and a half or less from Philadelphia to *the* "City by the Sea" as pleasant as human skill can make it.

Unfavorable as the country above Cape May was for travel, there was one circumstance of the early days which tended to make transit by water an occasion for apprehension, and rendered the worst "miry, boggy way" preferable to a route whereon the voyager had reason to look under every strange sail for the sinister visage of the sea-robber and pirate! The sixteenth century was an age of piracy, and as late as 1721 the Delaware was the scene of captures by the highwaymen of the ocean. Owing to its lack of naval and military strength, and to the reluctance of the Quakers to hang rascals, Philadelphia was a favorite place with Blackbeard and others of his kind, and the Delaware was chosen as a resort for repairs by many an outlaw vessel. In 1731, five men were hung as pirates, which was about the end of a bad bloody business in this part of the world. The pirates are said to have infested Sea Grove, and buried much money there, after which much digging and *conjuring* has been done, even in recent years.

Although most of the early inhabitants of Cape May were seafaring men, the Swedes among them were an agricultural people, and in time circumstances compelled the general cultivation of the soil. The colonies, near the close of the sixteenth century, were governed by the English lords of trade; every effort was made to prohibit manufactures and commerce in America. Still, whatever the wrongs of government, the natural resources of Cape May saved the settlement there from want. Parliament could not legislate the fish out of the Delaware, no lord of trade ever ate such oysters as fairly obstructed the sounds, no English park had half the game which swarmed in the woods and swamps, there was an abundance of wonderfully quick, fertile soil, easy of cultivation, and the sun never shone, even on an English king, as it beamed on the gardens and cornfields which, year by year, grew ever wider and still wider.

The primitive manufactures of Cape May, aside from lumber, were of a domestic nature, and were much encouraged by Jacob Spicer (2d); there was hardly anything that he would not take in exchange for goods. He advertised to receive, at the same time, a variety of produce, from a drove of cattle or sheep, "a thousand pounds of woolen stockings" for the army, or "a large quantity of mittens," to "a clam-shell formed in wampum, a yarn-thrum, a goose-quill, a horse-hair, a hog's-bristle, or a grain of mustard-seed, being," said he, "greatly desirous to encourage industry, as it is one of the most principal expedients, under the favor of Heaven, that can revive our drooping circumstances at this time of uncommon, but great and general burden." This was in 1756, during the French and Indian War,—the conflict wherein the colonists learned to "organize victory," and gained the confidence which made possible the triumph of the Revolution.

Cape May was fortunate in her early sons. Jacob Spicer (2d) was a statesman, a merchant, an economist; a man without conceit, he required in his own family the same reasonable diligence and thrift he recommended to others. There were twelve persons in his household, and such was his minutely systematic way of business that from his books and writings may be learned, even now, the details of their life. In Jacob Spicer's own house, under the superintendence of a tailor, tailoress, and shoemaker, the apparel of his family was made. The sons of this legislator and jurist were taught to cobble shoes, the girls to make clothing and knit. The Spicer boys, in 1757, were provided with "24 lbs. gray skin, @ 25d. per lb." to make them breeches and vests. This was deer-skin, and some of it was worn with the hair on. For the girls there was a provision of "striped linnen" and "linsey;" there was "a cloth vest" for one of the boys, and a "tammy quilt for Judith." Spicer estimated the girls to knit yearly, besides the other work they had to do, two hundred and twenty pairs of mittens, taking forty-four pounds of wool, to be spun by a hired woman in his house, in forty-four days. The mittens were worth "16d." (thirty-two cents) a pair at Cape May when finished, but sold at double the money at "York" and Albany. In one way and another, the premises of the Hon. Jacob Spicer must have been a lively place. Teetotalism had not been heard of at Cape May then, and under the head of "wets," the master of the house charges his family with using "52 gal. rum, 10 do. wine, and 2 bbls. cyder." As a merchant and magistrate, Spicer probably entertained many, and, in the unquestioned manner of his time, took care to "welcome the coming, speed the parting, guest."

The patient author, as he delves among these prosaic records of the past in the magnificent Centennial year of grace, 1876, remembers the scandals of his day, and pauses to heave a sigh, not for the leather breeches, linsey-woolsey, woolen mittens, rum and cider, of the Spicers, but for more of the conscientious good sense which made virtue, diligence, and economy the height of fashion, public spirit the pride of the citizen, and inflexible integrity the historical glory of the merchant and magistrate!

But with all the usefulness and sterling worth of Jacob Spicer (2d) notwithstanding he was for twenty-one years—nearly half his days—an officer and representative of his neighbors, still Aaron Leaming (2d), was the man the people of Cape May especially delighted to honor. He served them as their representative for thirty years: well educated for the times, of great natural good sense, very industrious, and, withal, somewhat aristocratic, no man was ever more highly honored by the county, and none, perhaps, better deserved the regard and confidence of his constituents. Neither Leaming nor Spicer were place-hunters, dependent upon local prejudice for recognition. Serving as colleagues in the assembly for a score of years or more, their ability and fidelity were made manifest, and together they were selected by the legislature for the responsible work of compiling the laws of the State. This they completed to the satisfaction of the public, and "Leaming and Spicer's Collection" is, to-day, a respected authority in New Jersey. Leaming was a great speculator in land, and yet found time to write copious "Memoirs," which remain a faithful transcript of the times in which he lived. Born in 1716, the son of Aaron Leamyeng, from Connecticut, a man who had worked his way against adverse circumstances to superior knowledge, large possessions, a Quaker faith, and public respect, Aaron Leaming (2d) maintained the honor of his family, filled with credit and dignity the important position assigned him, and died, much regretted, in 1783.

The Leamings, Goldens, Spicers, Stites, Stillwells, Willetts, Ludlams, Causons, Hands, Townsends, Youngs, Swains, Hughes, Garretsons, Hubbards, Mackeys, Godfreys, Reeves, and Weldons, Whilldens, or Whilldins, with others, were among the early and principal settlers of Cape May.

While history records the virtues of the early sons of the Cape, their prominence in seamanship, in commerce, in the halls of legislation, what shall be said of the women of the place and time? Theirs may have been a less conspicuous position, but many of them were of that class whose "children rise up and call them blessed;" diverse, yet equal, in domestic life they were accomplished in all good works, nor are we left without evidence of a bright intelligence, in many cases, to more endear them. Very early, the Quakers did much in West Jersey to modify and elevate the estimate of woman. In the library of Sarah Hall, of Salem and Alloway's Creek, Aaron Leaming the elder, as a boy, "very poor, helpless, and friendless," read law; the aged Quaker lady being herself "an eminent lawyer for those times." The student may fumble in vain among the dry leaves of court records, account books, and scattered memoranda, for the chronicle of great deeds by the mothers of Cape May; but while "like sire like son" has become a proverb, do we not know that the mother is equally the parent of the child, and that the men who have done honor to their native county learned, like Washington, their noblest lessons beside a mother's knee?

In an estimate of the resources, income, and expenditure of Cape May County, for 1758, made by Jacob Spicer (2nd), there is credit given the county for production of the "mitten article," to the value of five hundred pounds sterling. The manner in which the mitten trade, which, as thus appears, was quite a reward to the female industry of the County, was encouraged, is related in the following letter from Dr. Franklin to Benjamin Vaughan, dated Passy, July 26th, 1748, "on the benefits and evils of luxury:"

"The skipper of the shallop, employed between Cape May and Philadelphia, had done us some service, for which he refused to be paid. My wife understanding he had a daughter, sent her a present of a new fashioned cap. Three years afterward, this skipper being at my house with an old farmer of Cape May, his passenger, he mentioned the cap and how much his daughter had been pleased with it; 'but,' said he, 'it proved a dear cap to our congregation.' How so? 'When my daughter appeared with it at meeting, it was so much admired that all the girls resolved to get such caps from Philadelphia, and my wife and I computed that the whole would not have cost less than one hundred pounds.' 'True,' said the farmer, 'but you do not tell all the story. I think the cap was nevertheless an advantage to us, for it was the first thing that put our girls upon knitting worsted mittens for sale at Philadelphia, that they might have wherewithal to buy caps and ribbons there, and you know that that industry has continued, and is likely to continue and increase to a much greater value, and answer better purposes.' Upon the whole I was more reconciled to this little piece of luxury, since not only the girls were made happier by having fine caps, but Philadelphians by the supply of warm mittens."

The old times were trying times; hardly were the pioneers of Cape May settled in comparative comfort, before the entire country was plunged in the horrors of the French and Indian wars. Surrounded by the faithful Lenni Lenape, Cape May had no experience of the ruthless barbarities which were suffered elsewhere, but for many a year no one could tell when some French cruiser or Spanish privateer would break into the Delaware, and retaliate upon its defenseless shores the outrages Argall had imposed upon the French Acadians in 1613. New Jersey always cheerfully and with alacrity met the requisitions upon her for men and means; while the soldiers of Cape May faced a cruel foe, Jacob Spicer rallied the people to increased industry; "to meet the great demands of the time," he demanded "a thousand pounds of stockings," "for our men in the field;" faster than ever rolled the spinning-wheel, faster still flew the needles; even before the Revolution Cape May evinced a patriotic courage.

On the 1st of November, 1775, Jacob Spicer called a public meeting, "to do something for the country," but had to record his chagrin that only James Whillden, Jeremiah Hand, Thomas Leaming, and John Leonard attended. It was the era of doubt; the magic word, INDEPENDENCE, had not yet been uttered at Philadelphia,—the more honor to the ready few. Cape May sent Jesse Hand to Burlington as member of the Provincial Congress in 1775 and 1776. On the 21st of June, in the latter year, that body decided upon the formation of a new State Government. Hand was also a member of Council in 1779, and for three years afterward. Jesse Hand, Jacob Eldridge, and Matthew Whillden were the delegates sent from Cape May to attend the Convention at Trenton, on the second Tuesday of December, 1787, to ratify the Constitution of the United States; this was done by a unanimous vote on the 19th of the month, when the members of the Convention marched in solemn procession to the Court-House, where the act of ratification was publicly read. New Jersey was the third State to ratify the Constitution of the United States. By the Legislature of New Jersey Jesse Hand was made a member of the Committee of Public Safety, a most responsible and arduous position, but no one of those who served the cause of Independence, in a civil capacity, deserved better of his country.

Cape May has been noted for generations, as from natural causes, one of the best of beaches; the same peculiarities constitute it one of the most delightful driving places imaginable. Unequaled by nature, the beach road has been extended along shore, over Poverty Beach, away past the magnificent Cape May Lighthouse, past the beautiful cottages and comfortable hotels of Sea Grove, beyond the United States Signal Station, around the point, and for a perfect mile up the Delaware to the steamboat landing; from thence the straight inland road runs for three miles, over the turnpike, into Cape May City. Wherever the start be made, the seven miles round brings the rider to his door again. Hoof or wheel, it is the same good road, and all the way the ocean or the bay is constantly in view, and the surf can scarcely stir unheard.

But what have all these well-known facts to do with Jesse Hand and his offenses, he of ante-revolutionary fame? Well, the simple fact is, that gentleman and patriot utterly confounded, astonished, and disgusted his neighbors by his audacity in presuming to ride over the very route we have described, and others thereabout: the first man in the history of the world to traverse the roads and beaches of Cape May in the pretentious dignity and effeminate luxury of a top carriage. It was none of your modern affairs from Kimball, Brewster, or Rogers,

PUBLIC BATH-HOUSE, SEA GROVE.

but a solid, old-fashioned "chair," heavy enough, hard-riding enough; but what of that? Had not Aaron Leaming traveled on horseback to the Legislature? Had not everybody else ridden in *horse carts* year after year? And now Jesse Hand presumed upon a new and amazing fashion before their wondering eyes. History records no popular tumult, except of tongues, about the matter, but Jesse Hand never fully regained the regard of some people, and jealousy and distrust, like a curse, followed his new-fangled equipage; and though he and his generation are long since dead, yet the writer hath knowledge of traditions that, still drawn by attenuated and discouraged equines, a very Wandering Jew of vehicles, Jesse Hand's carriage still peregrinates, at a toilsome pace, the interminable, sandy, woodland roads of Jersey.

As to the part which Cape May took in the Revolution, Dr. Maurice Beesley, in his "Early History of Cape May," writes as follows: "In the contest of our forefathers for independence, nothing praiseworthy can be said of the other counties of the State that would not apply to Cape May. She was ever ready to meet the demands made upon her by the Legislature and the necessities of the times, whether that demand was for money or men. Being exposed, in having a lengthened water frontier, to the attacks and incursions of the enemy, it was necessary to keep in readiness a flotilla of boats and privateers, which were owned, armed, and manned by the people, and were successful in defending the coast against the British as well as refugees. Many prizes and prisoners were taken which stand announced in the papers of the day as creditable to the parties concerned. Acts of valor and daring might be related of this band of boatmen, which would not discredit the name of a Somers, or brush a laurel from the brow of their compatriots in arms. The women were formed into committees for the purpose of preparing clothing for the army, and acts of chivalry and fortitude were performed by them which were equally worthy of their fame and the cause they served. To record a single deserving act would do injustice to a part, and to give a place to all who signalized themselves would swell this sketch beyond its prescribed limits." Yet, on another page, the doctor cannot forbear telling the story of the devotion of Sarah, the sister of Captain Nicholas Stillwell, the young Mrs. Griffing. Captain Moses Griffing being a prisoner on the infamous and murderous "New Jersey Prison Ship," where the dying, the dead, the famished and famishing were promiscuously huddled together, Mrs. Griffing, "actuated by a heroism which woman's love alone can inspire," bravely made her way for a hundred and fifty miles through a most dangerous country, swarming with enemies, romantically resolved to see and rescue him, or die in the attempt. The devoted wife called at the camp of Washington by the way, who gave her in charge an English captain to exchange; she reached New York in safety, and finally persuaded Sir Henry Clinton to release her husband; the exchange was made, after a long and painful suspense, and the patriotic wife enjoyed the happiness she deserved.

And thus the men and women of "the time that tried men's souls" fought the battle of English liberty on the soil of America; and to-day the citizen of Britain may find in the extent and stability of his freedom abundant reason to rejoice in the result of a contest which, beginning in the colonies as successful revolution, culminated in the mother country in the achievements of liberal and progressive reform.

Notwithstanding the perturbations of war, or the changing policies of peace, Cape May prospered, and gradually enlarged its population. It was, however, no "Lotus Land," where the sdontaneous produce of the soil supported the inhabitants in corrupting sloth, to breathe an enervating air. Every ocean breeze of Cape May is an ethereal tonic, pure as the quintessence of the elixir of life. There are no long-prevailing, exhaustive extremes of torrid heat, and winter, comparatively brief, is only rigorous enough to destroy the germs of malaria, to superpurify the atmosphere with its frosts, and brace anew the vital powers of man and beast. The necessary pursuits of the pioneers were all manly, demanding hardihood, muscle, and courage; developing strength, heroism, and force of character.

There were about fifteen hundred people in Cape May County in 1758, with an estimated income of about twenty-two thousand dollars. When the war of 1812 began, the Cape had a population of three thousand five hundred persons, its commercial importance having increased in a greater degree. The final war with England was a naval contest; the interest of Cape May in such a struggle may be inferred. From first to last, in the various wars for freedom and independence, the waters in view from the towers of Sea Grove have been the scene of many naval conflicts. An interesting volume might be written of events when a British fleet lay constantly over against Henlopen; when they captured and burned the small craft of the bay, and in their launches cruised about, threatening to land and ravage the Cape.

What a romantic chapter the account of the watchful coast guard would make! And what an exciting scene it must have been, when the fast Yankee frigate, Alliance, then under Commodore Barry, fled out of the Delaware, to avoid a hopeless contest, and made her way to Rhode Island at the rate of fourteen knots an hour, running down the Speedwell and seizing two sloops of war, to fly back to the shores where her timbers grew, and land her wounded commander in the port of Boston! Then there was the first naval conflict of the Revolution, fought by the Hyder Ally, under the gallant Captain Barney, a privateer with four nine-pounder guns and one hundred and twenty-six men, which stole down from Philadelphia disguised as a merchantman, to attack the General Monk with eighteen nine-pounder guns and one hundred and fifty men on board. When the captain cried "Board!"

his men were to fire; when he cried "Fire!" they were to board. Alongside the Monk, Barry shouted his first command. The brave English crowd to repel boarders; the Hyder Ally rocks from stem to stern; everything that can carry a bullet explodes in the very faces of the foe. "Fire!"—in a flash the Yankee cutlasses are on the English deck! Doubly duped, twice tricked, the Monk surrendered: two-thirds of her crew were dead and wounded; but four were killed and fifteen hurt on board the privateer.

Another fight, turning the other way this time. A large American privateer, beset by a fleet of British launches, just off Cape May shore. A long fight, and a close one, until the vessel manœuvering too near the strand, strikes, and by and by goes to pieces in the breakers.

And so a book might be written of the waters around Cape May, as a scene of war and bloodshed. But to what good end? It could not prove that English hearts were cowardly, or that Americans were more than the world admits them to be. America won, in the last fight with England, because of finer modeled, better rigged, and more "handy" vessels; *and* because on those vessels, for the first time, long-range guns and cannon were supplied with "sights," and trained with the deadly accuracy of the rifle on the mark. It was the thunderbolt against the hail-storm; it was precision against mass; it was the rifle against the shot-gun; it was invention against routine; and science won, as it will forever in any fight. To-day, England sights her guns with telescopes; she clothes her warrior-ships in sevenfold steel; she buoys them with cork; she lights them with electricity; she drives them by steam, like avalanches; and by steam handles guns of eighty tons like toys, in the recesses of invulnerable turrets! Well, cannot the United States do as much? They have done, and are doing, better.

At the extreme point of Cape May, in the centre of Sea Grove beach, a tall spar bears aloft the flag of the American Union. Near by, a neat but peculiar building attracts the scrutiny of the observer. This is the United States Signal Station, and there keen-eyed vigilance watches and notes the skies, the clouds, the winds, the seas, and all the grand phenomena and minute signs of nature. On lofty mountains, amid deserts, by great lakes, everywhere throughout the territory of the United States, are similar posts of observation, and everywhere the same untiring watchfulness. The telegraphic wire links all these points together, and connects all with the central observatory at Washington.

It may be an overcast afternoon in September; nothing especially betokens danger, but vessel after vessel comes down the bay, catches sight of the station, and quietly passes behind the gigantic breakwater above Henlopen. An English ship sweeps down the coast, the cross of Britain bravely borne above her canvas; she too sights the station, and turns her helm, and bears sail, to gain, ere nightfall, sea-room and an offing. Night comes early, and with night the storm. The two great lights answer each other's glances across the bay, over seas which howl and show flashes of foam, like wolves snarling white-fanged in the tempestuous darkness! But the ships are safe, folded like sheep in a quiet place; for all day long the danger signal has been displayed, and they have learned to heed it; and that is an American idea, deserving fuller development, and worth more than the war-ships of the world.

There are three edifices most prominent at Sea Grove, the Light-House, the Signal Station, and the Pavilion: they typify the Nation and the Age; they actualize the beneficence of Popular Government, the philanthropy of Science, and the power of Moral Sentiment, in the sublimity of Religious Freedom: these, rather than batteries, armies, and navies, are the conquering forces of the future.

To show the critical and useful nature of the work done by the United States Signal Service, and as a matter of information, the following table is introduced; of the value of such statistics no well-informed person need to be advised.

The records of the United States Signal Service show the following figures for the three most prominent resorts on the New Jersey coast:

MEAN DAILY HUMIDITY.

	Cape May.	*Atlantic City.*	*Long Branch.*
July........	88.3	85.7	78.4
August........	78.8	79.0	77.4
September......	78.9	83.0	80.0
3 months........	82.0	82.6	78.6

MEAN DAILY TEMPERATURE.

	Cape May.	*Atlantic City.*	*Long Branch.*
July........	69 2	70.3	71.4
August........	68 8	69.5	69.9
September......	68.6	67.8	67.5
3 months........	68.9	69.2	69.6

Thus it is seen that Cape May is the coolest place along the coast, and as dry as Atlantic City.

The village of Cape May escaped the ravages of war. Once, in 1812, the Poictiers, a British line-of-battle ship, appeared off the place, and threatened it with bombardment unless it was supplied with water; the cheap ransom was paid at once, and the enemy sailed away. While the English fleet lay in Delaware Bay, in 1812, its officers managed to keep, so far as personalities went, on very good terms with the people of Cape May, and made "The Beach" what it is now, a place of healthful, free, and gay resort. The village of Cape May, though loyal, was hospitable, and the chronicles assure us that its amusements were shared by friends and foes together in the greatest amity, and that, when the fleet of Albion sailed away at last, more than one of the heroes and heroines of the time gave evidence of their faith, by obedience to the command, "Love your enemies."

In 1812, the present site of Cape May City was already the location of a considerable hamlet; even then popular as a place of resort in summer. "Cape Island" was purchased of Dr. Coxe, through his agents, by William Jacoks and Humphrey Hughes, in 1689—a tract of five hundred and forty-six acres, or more. Jacoks sold to Thomas Hand, and Randall Hewit bought an interest in the Island. Hand, Hewit, and Hughes held the property until 1700, and it was long cultivated and fertile land. But in the mean time the settlement increased, and the corn-fields were narrowed. In 1829, Watson, the annalist, visited Cape May City, "a village of about twenty houses," says he, "and the streets were very clean and grassy."

Very rapidly after the war of 1812 Cape May began to assume a distinctive character as a watering place, and its history from that time becomes modified accordingly. Gradually the fashions of Cape May have changed—are changing still, and not for the worse.

For an idea of the earlier methods of travel, and the ways and manners of sea-side visitors in the olden time, nothing can be better than the following, from *Lippincott's Magazine:* "Strange old sloops and bateaux used in those times to move slowly down the Delaware, bearing eager Philadelphians on pleasure bent. Other sojourners would drive miserably down in their dearborns, dragged by tired nags through the interminable sandy road from Camden. On the adoption of steam for navigation, a modest steamboat was conducted by Mr. Wilmon Whilldin, and cut its way down the long Delaware in what was deemed a fleet and stylish manner, greatly improving the prosperity of the place. The customs of those earlier times were very primitive and democratic. Large excursion-parties of gay girls and festive gentlemen would journey together, engaging the right to occupy Atlantic Hall, a desolate barn of a place, fifty feet square, whose proprietor was Mr. Hughes. Then, while the straggling villagers stared, these cargoes of mischief-makers would bear down upon the ocean, ducking and splashing in old suits of clothes brought in their carpet-sacks, and gathering the conditions of a fine appetite. The major-domo of Atlantic Hall, one Mackenzie, would send out to see what neighbor had a sheep to sell; the animal found, all the visitors of the male sex would turn to and help him dress it. Meantime, parties of foragers would go out among the farmers around ravaging the neighborhood for Indian corn. When the mutton was cooked and the corn boiled, an appetite would have accumulated sufficient to make these viands seem like the ambrosia of Olympus. Those were fine, heart-hold times, and when our predecessors at Cape May went down for a lark, they meant it and they had it. At night, when dead-tired after the fiddling and the contra-dances, the barn-like hall was partitioned off into two sleeping-rooms by a drapery of sheets. The maids slept tranquilly on one side the curtains, the lads on the other. Successive days brought other sports,—fishing in the clumsy boats, rides in hay-wagons over the deep white roads, the endless variety being supplied, after all, by the bathing, which was always the same and ever new. These primitive bivouacs were succeeded by a steady service of steamers on the Delaware and the erection of substantial and civilized hotels."

Thomas H. Hughes, Jonas C. Miller, R. S. Ludlam, and the Messrs. McMakin were among the first to erect large and commodious boarding-houses. Increasing custom demanded multiplied conveniences, and a host of varied places of entertainment grew up, from the small and modest restaurant to the monster hotel with its fifteen hundred guests at once. Meantime private cottages became numerous, the resident population enlarged, and a city was built up "where," says a writer in 1856, "a few years ago corn grew and verdure flourished." It would be a pleasant task to note the particulars of such a progress in full, and the reader could not fail of interesting information, but the work is left for another pen, or a future time. Material increase and prosperity is not the final test of development, and the scope of the present discussion demands attention to other and important matters.

§ Man is a religious being; the impulse to worship, an ineradicable instinct of his undying soul. Tyranny is the trait of the brute; it is the bestial element in man which offends against the prerogative of reason, and seeks, in intolerance, despotism over the spirit. Ambition and avarice enlarged their efforts to aggrandize themselves in the colonization of New Jersey; but, after all, the settlement of the State is found to be due, through persecution, to the love of liberty and the principles of religion.

"America," says an eminent historian, "was secured from bigotry by her welcome to every sect; each rallied round a truth, their collision could but eliminate error. The eclectic American mind struggled for universality while it asserted freedom. The Old World looked to the American Colonies for the benefit of commerce, for mines, for natural productions, but received revolutions,—the consequence of moral power." At Cape May, in as great a degree as in any other place, influences were early at work tending to hospitality of opinion and a broad and catholic spirit. Counting the whalemen as the pioneers of the county, Calvinism was the form of faith earliest introduced, but the Swedish Lutherans soon exerted an influence upon the community, and the Baptists and Quakers, not long after, were added as a powerful element. The English Church was strong on the shores to the west of the bay, where, for a time, the Reverend William Becket, an author and poet, held a broad parish. It had its adherents at Cape May also, but its connection with the monarchy, as an established Church, weakened its influence at the Revolutionary period.

The Baptists are said, in "Benedict's History of the Baptists," to have arrived at Cape May from England and formed a church as early as 1675. Johnson, in his sketch of Salem, says the same; but Dr. Beesley supposes a mistake in the date, as there is no record of a white population until 1685, none of a Baptist church until 1711. Elsewhere the doctor writes, "History throws no light on the original occupiers of the soil. Conjecture only can be consulted on the subject." It is quite probable that some of the Mennonist Baptists of Plockhoy's colony may have escaped to Cape May, from the spoliation of Carr, in 1664, with Swedes and Dutch from Christina and New Amstel,—refugees for the same cause.

The Baptists, "the scum of the Reformation," as they were called, were the democrats of the Protestant Church; the Calvinists aspired for theocracy, and made the Church dominant in the State; the Church of England took "submission" to royal prerogative as a "badge," and Luther taught that it was "a heathenish doctrine; that a wicked ruler may be deposed." But, plebeians themselves, the Baptists were consistent, and unflinchingly dealt with the relations of life, threatening an end to kingcraft, priestcraft, and feudalism. Hosts of the peasantry of Germany perished in the persecutions against the Baptists; arrogantly they were trodden under foot, and scorn and reproach heaped upon their memory. As might be expected, wherever the Baptists found shelter in America, in Rhode Island, Pennsylvania, New Jersey, they became a power, witnessing for independence, republicanism, and free religion.

One of the early pastors of the Baptists at Cape May was Nathaniel Jenkins, a Welshman, born in Cardiganshire in 1678. He arrived in America in 1710, and assumed his position in the church at Cape May in 1712. Mr. Jenkins was a man of character and ability, with fair education; from 1723 to 1733 he was a member of the Assembly; he was also a trustee in the Loan Office, and a local deputy and attorney of Governor Hamilton, in all of which positions he served with honor. Not long after the Baptist pastor became a legislator he had the opportunity of doing the state some service and distinguishing his principles. The emigrants from New England, accustomed to puritanical rigor, quite conscientiously strove for a long time to engraft their persecuting policy upon the institutions of New Jersey. When Mr. Jenkins was first a member, a bill was brought into the Assembly to punish such as denied the doctrine of the Trinity, the divinity of Christ, the inspiration of the Holy Scriptures, etc., etc. This the Baptist legislator opposed with all a Welshman's zeal and action. "I believe the doctrines in question," said he, "as much as the promoter of that ill-designed bill, but will never consent to oppose the opposers with law, or with any weapon save that of argument." "Accordingly, the bill was suppressed, to the great mortification of those who wanted to raise in New Jersey the spirit which so raged in New England."

The Baptist church was from six to seven miles north of Sea Grove, at what is now the district of Cape May Court-House; there Mr. Jenkins died and there is his grave.

The early presence and work of the Baptists at Cape May perhaps left fewer to adhere to the Quakers. There were a number of Friends at the Cape in the early days, but they never became as numerous as in Cumberland and other counties. Neither the Friends nor any others have been persecuted at Cape May. Quakers generally went where they were *not* wanted, but needed; sometimes their peculiar principles subjected them to loss, even at Cape May, but no persecution appears to have been intended. Thomas Leaming, who came to Cape May in 1692, and became a whaleman, and then a farmer, seems to have been a Quaker; among other things he records: "In 1706, I built my house. Samuel Matthews took a horse from me worth £7, because I could not train." "Training" would have prevented the levy, and paying the fine for contempt would have saved the horse from sale; Quaker principles forbade one and the other; it was hard upon Thomas Leaming, but what could Samuel Matthews do with a Christian who would not fight—nor swear!

Within the memory of the elder people of this generation, a Quaker meeting-house stood in the northern part of Cape May County, and there the tradition is that year after year, every First-day, two old Quakers got together, and silently sat out the hours; furthermore, these Friends were not friendly, not on speaking terms, and so spake not at all. By and by one of the old men died, and then the survivor sat alone, scarcely more solitary, no whit more silent, until at last he too came no more. But the part the Quakers took in founding Cape May County has not been without a permanent effect for good,—and there are men to-day everywhere who, could they but learn to hold their tongues as faithfully as the two in the above story, "the world would be the better for it!"

Pre-eminently, Calvinism has appealed to the human intellect. The Democratic State, Free Church, and Common School arose together. The Church which invoked thought, as a co-worker with zeal and faith, gave guarantees to progress; the Antinomians of Massachusetts advanced beyond Geneva, and in Connecticut, where Calvinism enjoyed a hundred years of peace, Massachusetts was left behind. There religious pride was forgotten; predestination was less considered than philanthropy; persecution was abandoned, and reason and charity were made the basis of law. "Virtue," said the great Connecticut Presbyterian divine, Jonathan Edwards, "consists in universal love." From the churches of Connecticut were drawn the men and women who planted Presbyterianism at Cape May; there also freedom and peace favored the finest developments, and the influence of Calvinism may be recognized in the stability, thoroughness, and intelligence which have characterized the people.

BATH-HOUSE, SEA GROVE.

The first Presbyterian church in Cape May County was established at Cold Spring, two miles to the north of Sea Grove; its earliest chronicles have disappeared, but it is recorded that the first minister was the Rev. John Bradner, a native of Scotland. Mr. Bradner was a candidate for the ministry when invited to Cold Spring, but Rev. Allen H. Brown, in his "Outline History of the Presbyterian Church in West South Jersey," says he had no authority to preach, and it marks the unorganized state of Church affairs at the time, that Messrs. Davis, Hampton, and Henry, the three nearest ministers, took the responsibility of examining and licensing him in March, 1714, he being ordained May 6th, 1715. In 1721 Mr. Bradner was removed to Goshen, Orange County, New York, still keeping, however, his connection with the Presbytery of Philadelphia. Mr. Bradner died before September, 1733. The estate now occupied as the parsonage, consisting of some two hundred acres, was conveyed by him, for the use of the pastor of this Church, to Humphrey Hughes, George Hand, John Parsons, Joseph Weldon (Whilldin), James Spicer, and twenty-seven others. Mr. Bradner was succeeded by Rev. Samuel Finley, who, as a resident at Cape May, often officiated, though not settled there. Mr. Finley was distinguished for learning and personal holiness; the great revival among the Presbyterian and Baptist churches from 1740 to 1743 was regarded as, in a large degree, God's blessing upon his labors; in 1761 he became President of Princeton College; he was made a doctor of divinity, by the University of Glasgow, in 1763, and, after an active and useful life, died in remarkable peace and happiness, at Philadelphia, July 16th, 1766, being then fifty-one years of age.

From 1721 to 1751 the Cold Spring Church had no settled minister; Messrs. Beatty, Dean, Davenport, and others, were a temporary supply. The Rev. Daniel Lawrence was at last installed, June 20th, 1754. Of his ministry little is known, except that in addition to his labors at Cape May he was often at the Forks of Brandywine, and, in 1755, went to preach at "New England over the mountains." Mr. Lawrence ministered to the Cold Spring Church twelve years; he died in 1766; his grave is at Cold Spring. After the decease of Mr. Lawrence, the Rev. John Brainerd supplied the Cold Spring pulpit in 1769 and 1770, and there is a report that a Mr. Schenck, a progenitor of the Hon. Robert Schenck, preached at Cape May, probably about this time.

Mr. James Watt was the next minister; the tombstone records his death November 19th, 1789, aged 46 years. Mr. Watts is said to have been a man remarkable for disinterested kindness, integrity, and ability; he was of the First Presbytery of Philadelphia, and represented that body in the General Assembly of 1789.

The sombre manners of some of the stern New England forefathers gave reason for an accusation trite as untrue, that Presbyterians were always an austere, sour, morose set of ascetics; the biographical anecdotes of the Rev. Mr. Watt might be quoted in refutal. Like the apostle Peter, Mr. Watt was much inclined to "go a-fishing," and of all fish the devil-fish was the one he most delighted to pursue. "The devil" is often a large, powerful fish, its capture rough sport. On one occasion, while accompanied by two other clergymen, Mr. Watt harpooned a devil-fish in Delaware Bay, so large and strong that it rapidly drew the boat toward the sea. Amid the apprehension felt by all, but especially by his guests, Dr. Watt, as he was familiarly called, broke out in hearty laughter. He assured his companions that he could not conceal his amusement at the idea of three clergymen of the orthodox Church being run away with by the Devil!

Mr. Watt was succeeded by the Rev. Abijah Davis, of whom there is no record. The Rev. David Edwards followed him, dying in 1813. After 1808, the church-record has been preserved. The Rev. Isaac A. Ogden was installed 1817; he resigned and went West in 1825. His successor was the Rev. Alvin H. Parker, installed June 19th, 1825, on the occasion of the first meeting of the Presbytery at Cold Spring. The elders composing the session were Matthew Whillden (Whilldin), John Stites, Jacob Foster, Jesse Hughes, and Jacob R. Hughes. Moses Williamson was ordained and installed at Cold Spring in 1831; he founded a successful academy there. Under his ministry the church prospered; he resigned in 1872, being now an honored resident of Cape May City.

The pastor at Cold Spring is the Rev. Thomas S. Dewing, to whom, with the Rev. Dr. Alfred E. Nevins, thanks are due for items of this history. Mr. Dewing began his labors October 1st, 1873, and was installed May 6th, 1874. The Cold Spring Church has two hundred members; the Sabbath-school two hundred and fifty scholars; a chapel has been built near the Cape, and the church improved, the means tberefor being derived from a legacy by Hon. Matthew Marcy.

The Presbyterian church at Cape Island was erected in 1845, as "the visitors' church;" there the Rev. Mr. Williamson, before mentioned, preached on Sabbath and Tuesday evenings, until 1851, when the Cape Island Presbyterian Church was organized. The Rev. E. P. Shields is now in charge, and under his "diligent culture" and "judicious oversight," says Dr. Nevins, the society is prosperous. There is also an Episcopal, a Methodist, and a Baptist church at Cape May City.

In the Cold Spring church-yard, and in another burying-place near the bay, above the steamboat landing, are entombed the ashes of many of the pioneers of Cape May, and of the generations which followed them. The tombstone is the only record of some who, once active and conspicuous, are now no more regarded; their names, their memory, obsolete, except among the venerable few.

ALEXANDER WHILLDIN, FOUNDER OF SEA GROVE.

But an enlightened faith dwells not in tombs, and recognizes death only as an incident of life; they whose bodies went down into the grave at Cape May came to their earthly consummation in a land where the hope of a happy immortality was part of the common creed. Strong in religious faith, upheld by a consciousness of spiritual things, the hour of their departure was to many of them an hour of triumph. One and all, they lived—they died; their example still remains: in high or low degree they filled the sphere they found, "whereunto they were appointed." The soul scorns the history that ends at the grave; as we stand amid the trampled dust of by-gone myriads, it lifts its voice within, to assert the presence of the angelic hosts and proclaim over all the just and loving providence of God.

Considering thus briefly the history of the churches of Cape May, regarding especially the Presbyterian organization, it is noticeable that while Calvinism has been influential, it has by no means been the sole creed of the people; the author follows the record of Presbyterianism at greatest length, because the past history of that church and the recent action of Presbyterians at Sea Grove are strikingly pertinent to his argument, that freedom and equality are the safe basis of material and spiritual progress.

Society, as some imagine, depends upon despotism, and religion they think a tender plant, thriving best in the shadow of a throne, hedged by bayonets; grafted, at least, upon some constitution, and guarded by facile courts. The Presbyterian Church is a free, a self-governing republican church. New Jersey has been democratic in the extreme, and absolutely tolerant. At Cape May Presbyterianism has had freedom, continuity, scope, and time, without isolation; it is fair to accept the outgrowth as a test of democratic republicanism and of the tendency of Calvinism in the United States. To learn this requires, in addition to a survey of the past, an observation of the novel yet characteristic developments of the present.

§ In passing from that which has become historical in relation to the Presbyterians of Cape May to an observation of the present, one thing may be remarked of peculiar interest and significance: the faith of the forefathers has, as it were, become hereditary; the names of prominent Presbyterians to-day are those found in the old church chronicles and traditions. Thus, Joseph Weldon (Whilldon, Whillden, or Whilldin) was one of the original trustees of Cold Spring Church. Matthew Whillden (Whilldin), with John Stites and Jesse and Jacob R. Hughes, were elders of the session in 1825, and to-day Alexander Whilldin, after serving his church for a full generation as an elder, is now President of the Sea Grove Association. Of the above officers of the church, Mr. John Stites and Matthew Whilldon each held the position of "active and ruling elders" for fifty years or more; they were contemporaries, and their terms of office were nearly coincident.

The organization over which Mr. Alexander Whilldin presides, having both secular and religious purposes, is Presbyterian in its antecedents and affiliations, though not exclusively sectarian in its constituency and designs, yet its work has been one of the most striking manifestations of character and tendency given by Presbyterians for many years; and hence the value of the history of the Association, and its force as evidence in establishing the assumptions which are embodied in the first paragraphs of this work, and which are the conclusion of its argument.

Following the course of our narrative and the discussion together, the history of Sea Grove becomes repuisite, and immediately in order. No account of the Sea Grove Association, its origin and operations, can be at all complete without some sketch, more or less circumstantial, of the gentleman who presides over the business of that corporation; hence the necessity of reference to him in the succeeding paragraphs.

The founder of Sea Grove is a native of Philadelphia, having been born in that city in 1808; his father was a sea-captain, and a native of New Jersey. In 1812, leaving France on a return voyage to this country, he never reached our shores, no tidings of his fate ever coming to relieve the suspense of the bereaved family. This sad event left Alexander Whilldin an orphan at the early age of four years. The widow with her son and two daughters left Philadelphia, and went to reside at the old homestead in Cape May County.

There, on the old farm near the Court-House, Alexander lived for twelve years, receiving only the meagre education that the country school-house of that day could give. In his sixteenth year he returned to Philadelphia and entered a store, where, without grumbling, he performed the duties of youngest clerk, including making the fires, sweeping the store, running of errands, and other things too often counted as drudgery nowadays. He was not too proud to work, and he worked earnestly, industriously, faithfully. He remained here as clerk eight years, rising from one position to another, gaining the confidence of his employer and of all about him. In 1832 he started business for himself, as a commission merchant in cotton and wool, the first year with a partner who brought in needed capital, and afterwards alone.

His career as a business man now began in earnest. He soon developed those traits which mark the solid man of business wherever you find him. He was prudent, sound in judgment, courteous, self-reliant, industrious, and of indomitable energy and persistence. He at once gave proof of great executive ability, and of capacity to direct extensive and complicated affairs. With such a power at the helm, his business rapidly grew to large proportions; and although at one time embarrassment surrounded him, his native resources of energy, sagacity, and superior judgment enabled him finally to extricate himself honorably, to meet every obligation he had assumed, and to build himself up on the experience of his trials upon a broader and surer basis than ever.

The peculiar talents of such a merchant could not of course remain the exclusive possession of his own large business. Mr. Whilldin was sought in commercial and financial circles, and was for many years President of the American Life Insurance Company of Philadelphia, and prominently interested in the management of other pecuniary trusts. His philanthropic sympathies, known generosity, and the personal interest he has always taken in educational, benevolent, and religious enterprises have made him prominent also in many noble public charities of Philadelphia; his upright character, and wisdom in counsel, making him invaluable as a leader in his own church denomination, and very efficient as a manager in the American Sunday-school Union, Presbyterian Hospital, and other worthy educational and philanthropic institutions.

A truthful likeness of Mr. Whilldin is included in the illustrations of this work. Although nearing his seventieth year after a laborious life, he enjoys the reward of ever temperate habits in an eye as clear, a step as elastic, and a mind as vigorous as most men of fifty. He is still actively at the head of his extended business, which is conducted in company with his three sons, and remains, as he has been for a full half-century, a respected and useful citizen of the great city where he was born. Though so long a resident of Philadelphia, the experienced merchant has never outworn his fondness for the scenes of his boyhood, or failed to appreciate the advantages of an annual sojourn at the old familiar sea-side places. It has been for many years his delight to escape from the cares of business, and seek beside the waters of Cape May the recreation which nowhere else seemed as grateful and complete. For half a century, except one season in Europe, he has been there every summer.

Not only as a visitor to Cape May, but as the holder of considerable real estate there, Mr. Whilldin has watched with interest the growth and peculiar prosperity of the county; desirous, as a philanthropist, that all should enjoy and be benefited by the natural peculiarities of the place, he has seen, with regret, the increase of a bad fashion which renders the season for rest and health-giving resort to nature but a wearying round of dissipation. "More than twenty years ago," says the founder of Sea Grove, "I had this subject under consideration." Many beside Mr. Whilldin had long deplored watering-place extravagance, and several denominations had established quiet places of congenial resort for their members, but none existed among Presbyterians. Providentially, as some of his friends declare, Mr. Whilldin was in possession of a most convenient location, whose great but long-reserved natural advantages invited occupation; besides, he had the courage, the means, and the influence, to make successful whatever he considered it his duty to undertake.

The site of Sea Grove was purchased of "The West New Jersey Society in England," by Jonathan Pyne the elder, through Jeremiah Basse as agent; being inherited by Jonathan Pyne (2d) and Abigail Pyne, it was deeded by them and Robert Courtney, Abigail Pyne's husband, to Henry Stites, in 1712. The property remained in the Stites family until the marriage of Jane G. Stites with Alexander Whilldin, in 1836, and was by them conveyed to the Sea Grove Association, March 15th, 1875, having been in Presbyterian hands one hundred and sixty-three years.

"I have come," said Mr. Whilldin, "to consider it the providence of God that we have been led to retain this Presbyterian ground all these years, to become a subject of special consecration at last." To him it seemed little less than desecration to appropriate the place he loved to the use for which nature had pre-eminently fitted it—that of a superior sea-side resort—if it must be done in the ordinary manner. Yet it seemed a pity so fine a place as Sea Grove should benefit so few, especially when scores of thousands in the great cities not far away needed every summer the comfort and help of the ocean air, and yet found themselves repelled and excluded from most popular resorts by the crowding, the confusion, the mad revelry, and recklessness which more and more characterized them.

Under the circumstances, Mr. Whilldin took counsel in the first place, as has been stated, of his own thoughts and inspirations, for some time considering the matter; then, like a wise and practical man, he conferred with his wife. "We," said he, "laid the matter before God;" and then, feeling as if Heaven intended to bless the work, the Presbyterian man of business conferred with his brethren. His suggestions were generally approved, and it was decided to utilize the choice location at the point of Cape May, and "*furnish a Moral and Religious Seaside Home for the glory of God and the welfare of Man, where he may be refreshed and invigorated body and soul, and better fitted for the highest and noblest duties of life.*"

In furtherance of this object an organization was effected the 18th of February, 1875, which was chartered, with liberal franchises, by the Senate and General Assembly of the State of New Jersey the same year.

This corporation was styled the "Sea Grove Association," and its Board of Directors was made to consist of Alexander Whilldin, Dr. V. M. D. Marcy, Downs Edmunds, J. Newton Walker, and John Wanamaker.

Section 6th of the charter vests the Board of Directors with the power of regulation and control in the following terms:

"6. *And be it enacted,* That a majority of the directors for the time being shall form a board for the transaction of the business of said corporation, and shall have power to make such by-laws, ordinances, and regulations as shall seem necessary and convenient for the management and disposition of the stock, effects, and concerns of the said corporation, and for the purpose of restricting nuisances and of compelling a uniform system of improvements; the said company are hereby authorized and invested with power to incorporate into any deed or conveyance made by them, whether in fee simple or otherwise, a clause or condition forbidding the sale upon the premises of any spirituous or intoxicating liquors, and to require of any grantee of said company to make and maintain such style and character of improvement on said lots so conveyed, or on the streets fronting thereon, as to the said company may seem best for securing a uniform system of development and improvement throughout the said settlement; and the board of directors of said company shall have the power to appoint such peace officers as they may deem necessary for the purpose of keeping order on the premises, which officers shall be paid by the said company, but shall have when on duty the same power and authority and immunities which constables and other peace officers under the laws of this State possess or enjoy when on duty as such, and they shall have the same power to enforce obedience to any rule and regulation of said corporation for the preservation of quiet and good order on the premises of said corporation and their grantees; *provided*, that such by-laws or regulations are not contrary to the laws or constitution of the United States or of this State."

Subsequently the officers of the Association were elected, with Alexander Whilldin, President and Treasurer, 20 South Front Street, Philadelphia; J. C. Sidney, Secretary, 204 South Fifth Street, Philadelphia; Downs Edmunds, Assistant Secretary, Cape May Point.

The Directors of the Sea Grove Association adopted a series of by-laws and regulations which provided for the systematic and business-like conduct of its affairs, according to the terms of its charter. Of these regulations the 12th, 13th, and 15th are special and significant in character, and by their nature or general interest, and are therefore here inserted:

"12. All buildings and other improvements will be subject to the approval of the directors or to an agent appointed for that purpose by them.

"13. All deeds will contain a clause for the purpose of restricting nuisances and of compelling a uniform system of improvement, forbidding the sale or keeping for sale any spirituous or intoxicating liquors, and generally providing for the submission to such rules and regulations as the Board may from time to time direct. Neither the holder of a lot, stockholder, or other person shall permit any amusement or act inconsistent with the character of the place and the objects of the Association, as set forth in the charter. * * * * * * * *

"15. The Pavilion is intended for religious or other meetings. Parties desiring to occupy it will not be permitted to do so without the authority in writing of the President, or in his absence of the majority of the Board, who shall first ascertain the character of the intended meeting, refusing the use thereof to all such as are not compatible with the objects and purposes of the Association."

The stock of the company being taken up at once, operations began, and were pushed with great energy. After the manner of the Calvinist Puritans of New England long ago, the first structure in the new settlement was an edifice for the purposes of education and religion. "The Pavilion," though vastly different from the comfortless churches in which the Pilgrims delighted, was yet the creation of the same spirit, though working under vastly different circumstances. As a building, it is well adapted to the purpose for which it has been constructed. In whatever respect it may fall short of the too great splendor of some city sanctuaries, it has one excellence in which many costly churches are deficient: as seen by the view on another page, it admits of perfeet ventilation.

Completed late in the spring of 1875, the history of the Sea Grove Pavilion is evidently brief; still, as the centre of Sea Grove enterprise, it has already attracted much attention, and been the scene of several memorable gatherings. For the ensuing record and description of the first season of Pavilion meetings the author is indebted to the Rev. Alfred E. Nevins, D.D., the superintendent and friendly manager of the services. Though a distinguished array of clerical talent of various denominations was always available through the season, yet to the supervision and care of Dr. Nevins was due much of the regularity and success of those assemblies. In connection with the regular Pavilion services on Sunday, a Sabbath-school was organized, and conducted with decided success by Mr. S. E. Hughes, a member of the Methodist church.

Recounting the already stated objects of the Sea Grove Association, Dr. Nevins proceeds to remark: "During the season of 1875, it was truly gratifying to see how this 'object' was kept in view, appreciated, and carried out. Early in July, in the attractive structure set apart for Divine service, and standing where but a few months before were seen the dense and dark forest, hundreds of visitors were summoned from the commodious hotel and handsome cottages by the clear and sweet tones of the bell, ringing through the grove and along the beach, to engage in the worship of their common Creator and Redeemer. And on every succeeding Sabbath was the same invitation given and accepted. A pleasing spectacle it was, on such occasions, to behold those who, though differing in some points of faith, yet agreed in the essential elements of our holy religion, the veteran and the child, the stranger and the familiar friend, all mingling their voices and uniting their hearts in praise and prayer to the God and Father of our Lord

SEA GROVE HOUSE, ENLARGED. 1876.

Jesus Christ, and listening with eager ear and ardent interest to the exposition of His most excellent Word. Still more pleasant, if possible, was it to see several hundreds of representative ministers and laymen of the various evangelical denominations, who had been invited to convene for the consideration of great moral and religious subjects, on the 25th of August, engage for several days in the noble Pavilion in earnest devotions and discussions, and then on the holy Sabbath unite in celebrating the love of Him who 'died for our sins, according to the Scriptures,' who prays that His followers 'all may be one,' and who has gone to prepare for them a place where they shall dwell together in blissful fellowship through an endless existence. There is something in the magnitude and grandeur of the ocean, as it is gazed upon, a great symbol of eternity, to overwhelm the mind, and cast minor matters into the shade. The very sight of it, in this view, tends to magnify the essentials of Christianity and to minify its circumstantials. And this effect was evidently realized in no small degree by the visitors at Sea Grove. May the flame that was kindled on the 'shore'—a spot which, in another land, Jesus so much delighted to frequent—send its light and heat throughout the country and the world!

"'One sole baptismal sign,
One Lord, below, above.
Zion, one faith is thine,
One only watchword—Love.
From different temples though it rise,
One song ascendeth to the skies.

"'Oh, why should they who love
One Gospel to unfold,
Who seek one home above,
On earth be strange and cold?
Why, subjects of the Prince of Peace,
In strife abide, and bitterness?

"'Head of the church beneath,
The catholic—the true—
On all her members breathe—
Her broken frame renew!
Then shall Thy perfect will be done,
When Christians love and live as one.'"

While the Pavilion was going up, a heavy force of men were grading the streets, avenues, and boulevards of Sea Grove. In the "Bird's-eye View of Sea Grove" which illustrates this book, the plan appears as laid out by Mr. Sidney, the architect. At the same time, the foundations of the "Sea Grove House" were laid, and the building pushed with great energy, being ready for use and thronged with hundreds of guests the same season. Simultaneously, many cottages arose here and there, all neat and attractive, and some ornately elegant; notably that of the Whilldin family, illustrated, with the original Sea Grove House, in the view of "Atlantic Beach," from a photograph taken upon the completion of the first buildings of Sea Grove. Besides, as early examples of fitness and characteristic good taste, might be mentioned the cottage of John Wanamaker; one for Mr. Stockton, an Episcopalian divine; several built on account of J. Newton Walker, M.D.; those constructed by Mr. Hughes, etc., as shown in the bird's-eye view.

Episcopalians, Lutherans, Methodists, and Christians of other denominations were already owners of the soil, and began to build; besides, people, not members of any church, were drawn by the promise of order and morality to seek a resting-place in the borders of the new, fast-growing town. The frontispiece represents Cape May Point, the site of Sea Grove, as it appeared in 1776; the picture of "Sea Grove Beach by Moonlight" is a sketch of the same locality, from a different point of view, just before the lighthouse was moved inland, the constant action of the sea having worn away the low bluff until the tower would soon have been in danger of a fall into the encroaching waves. These engravings are essentially accurate, except that in the original paintings, from which both of them were copied, the artist, Mr. Charles W. Knapp, of Philadelphia, well known by his fine authentic American landscapes, though reproducing faithfully and beautifully from older sketches and various data the natural features of each scene, has introduced, for artistic effect, more figures of men and women than often gathered on "Barren Beach" in those days. The lighthouse being moved back in 1847, the scene remained without much change until 1875, when the improvements of Sea Grove began.

The "Bird's-eye View" is reduced from a design by Armitage, of Mr. Sidney's office, and gives the appearance of Sea Grove in the spring of 1876. Studied thus in connection and contrast, these pictures are more expressive than any words the writer can command. The other views in this volume, being from sketches by David B. Gulick, of New York, are, of course, reliable pictures; they present artistically the actual features in a peculiar landscape, and, as will soon be seen, are significant of the remarkable influence of changing phases of religious sentiment upon general progress. Let the reader look at the picture of the gateway of Sea Grove, at the view of Lake Lily, and at the architecture of all the buildings in the various scenes, and then compare the liberality, taste, and good sense of Presbyterians to-day, with the temper manifested by the conscientiously ugly and uncomfortable "meeting-houses" of New England Calvinists two hundred years ago!

§"A sea-side resort," say the Directors of the Sea Grove Association, in one of their publications, "is generally associated in the mind with lavish display, extravagant living, dissipation, and consequent expense, to be regretted when the apparent pleasure is past and gone. To families of quiet habits, and who visit a summer resort, even where expense is no object, the glitter and show do not compensate for the health lost or for bad habits formed, especially by the young.

The fashionable hotel at a watering place may afford at enormous prices some luxuries and some exciting amusements, but attending these are generally small, inconvenient rooms, dissipation of every kind, a mixed, often immoral, company, the irreverent element preponderating over the moral and religious." In view of these facts, they explain that their enterprise has been undertaken "with the idea of affording a sea-side resort, and sea-side homes, with their economies and pleasures, as well as the influences arising from a religious sentiment, good order, and a freedom from all dissipation attending the merely fashionable watering places."

To encourage home life and influences at the sea-side, the greatest inducements are extended to those who build at Sea Grove. Aside from all other advantages of ready-made and perfectly graded streets, etc., each builder of a cottage will be entitled to a free pass over the West Jersey Railroad for one, two, or three years, agreeably to cost of improvement; and all materials will be carried at a reduced rate. The Association, to facilitate transit from Cape May City and the steamboat landing to Sea Grove, have constructed a horse railroad between those places. Good-sized lots at Sea Grove have been put at moderate prices, and, to prevent monopolizing speculation, none are sold except to those who agree to build within three years. The observer can detect neither overreaching greed or insane fanaticism in the developments of Sea Grove; the enterprise is no crusade, no pilgrimage to some "holy" but unhealthy sacred place, at the command of superstition.

Presbyterians to-day expect God's blessing of health only as they conform to natural laws, the dictates of sanitary science, and good sense. Speaking of the site of their enterprise, the Directors announce: "The land is sufficiently rolling to afford good drainage in every direction, and there are many building-sites rising twenty-five feet above the level of the ocean. There is no swamp on the tract, and the whole plot is available for building purposes. Water for drinking and culinary purposes, of the purest quality, is obtained on any part of the ground at a depth of sixteen feet from the surface."

Hospitable religion, broad boulevards, perfect drainage, pure, plentiful water, hygienic living; this is the Presbyterian programme to-day. Not very long ago, moody, mistaken "saints," of varied sects, counted religion, or the madness they called such, godliness enough, leaving cleanliness and care for the body to be regarded almost as a vice; herein is evident improvement. Progress involves no shifting of the grounds of principle, no change in the immutable basis of truth; it is a matter of perception and receptivity. It is mankind that is "converted from the error of its way," to grow in intelligence, in morals, in spiritual unfolding, to the measure of a perfect life!

Peculiar in its origin, remarkable in its development, striking in its results, as a work of high civilization, as an index of progress, Sea Grove commands attention by the liberality, skill, and judgment everywhere evinced, but is equally a display of good taste, a substantial recognition of the claims of the beautiful. Calling to mind the "crop-eared boors" of Marston Moor, the "Roundheads" of Cromwell's army, the parliament of "Praise God Barebones," the grim Puritans of Salem and Boston, the Pilgrim Fathers, "the Saints" of New Haven, the Covenanters of Scotland, and the early Presbyterians of New Jersey, with all the stern dogmatists of a persecuted denomination whose members were accused of regarding propriety and comeliness as a "wile of the wicked one," how strange, how inconsistent seem the works and ways of their lineal descendants in the spirit—the inheritors of the faith, the Presbyterians of to-day. Yet, whatever may, at first thought, seem to be the case, Sea Grove is a coherent outgrowth of Geneva, and Calvinism as much at home there as it was with the democratic and catholic Pilgrim. Fathers aboard the Mayflower and at Plymouth, or amid the privations, gloom, austerity, and exclusiveness of the persecuting Puritans during the first years of Salem and Boston.

To make it still more plain that the principles, good manners, and morals of the elder generations of Calvinists are essentially preserved, and their foibles alone omitted, by the people of Sea Grove, the following, from the pen of an experienced, observing, orthodox minister, is here added: "Throughout the season a bright, cheerful, and sociable spirit prevailed. Innocent and agreeable amusements abounded. Cultured and friendly intercourse was cherished. Guests, without any constraint to do so, had an opportunity of attending family worship morning and evening, as well as public Divine service on the Sabbath. Nothing occurred of any kind to mar the pleasure of the visitors from North, South, East, and West, and every sign indicated a brilliant and useful future for Sea Grove. It was evident to all who visited the place for a day, a week, or a month, that it is just the resort that is needed, one where fashion and dissipation do not hold sway, where extravagance finds no sphere for display, where guests, without an affected pretense of piety or of devotional services, may enjoy the means of grace to which the inmates of Christian homes are accustomed in their church relations, and where, whilst religious advantages are supplied and cherished, everything like sectarianism and bigotry is eschewed. It was a joy to the writer that, during a visit of a number of weeks at Sea Grove, whilst witnessing much cheerful enjoyment, and sharing in it, he neither saw a card or glass of liquor, nor heard a profane word. And this was peculiarly gratifying, as so many young persons were present, who could not but be benefited by so much exemption from evil influences, while under the power of others of an opposite character, genial, cheering, manly, and eminently salutary."

§ It has been stated that not only Presbyterians, but Episcopalians,

Lutherans, Methodists, and others were attracted to Sea Grove. To show how some of these people regard the place and its arrangements leads the author directly to the point of his perhaps too long and circumstantial argument. The subjoined paragraphs are from an Episcopalian in training and by long affiliation a liberal man and a reformer, one familiar with great enterprises, and to whom some of the most remarkable acts of the Senate of the United States owe their conception. This gentleman made a careful examination of the affairs of Sea Grove at an early date, and in a letter from thence to a friend, but for publication, he wrote,—

"Another visit to this delightful spot demonstrated the fact that the success of the enterprise is assured.

"Although somewhat surprised at the rapid advancement of the work as it appeared a few weeks ago, the sight presented last Saturday seemed quite bewildering.

"The place, considering the brief interval, had assumed the air of a lively little village, with evidences on all sides of the greatest activity and healthy progress. Cottages are springing up as if under the inspiration of magic, the commodious hotel recently begun is now open, the beautiful wide avenues are being graded and graveled, the sidewalks gently elevated above the smooth, level drives, and the busy workmen finishing their labors give hopeful note of preparation for the coming season.

"The excursionists last Saturday appeared to enjoy the visit very much; after taking a bird's-eye view of Sea Grove from the lofty steeple of the Pavilion, they strolled off along the beach, and visited the numerous objects of interest, as the clear fresh-water lake, the new hotel, the cottages, and grounds.

"A sign of the progressive times is the fact that men of large experience, keen sagacity, and ample means are attracted to South Jersey partly from its superior natural advantages, and, in some measure, from the moral tone and growing sentiments of the people upon the vital questions of temperance and prohibition, without which no community can reach the highest degree of moral development and material wealth.

"The Sea Grove Association, composed of such men as Messrs. Alexander Whilldin, John Wanamaker, J. C. Sidney, and others of like earnestness and capacity, recognizing these principles, has founded this new settlement on the basis of morality, religion, and temperance, and procured such legislation as will effectually banish, within the corporate limits, the sale and traffic of intoxicating liquors. Built on this superstructure, with its natural advantage of position, health, accessibility, moral tone, and religious sentiments, the future of Sea Grove is assured.

"The temperance feature is one which your valuable journal cannot too highly commend. The beneficent result which will soon be apparent from the practical workings of the prohibitory provision will furnish you additional arguments and illustrations to continue the manly fight made in Pennsylvania against the rum traffic and in favor of prohibition throughout the State.

"The party returned to the city about eight o'clock without accident, seemingly well pleased with the day's enjoyment."

But all the world are not Presbyterians, all are not Episcopalians; some are not members of any church, and yet are people of discrimination, at least in secular matters. John Calvin, in Geneva, could not raise himself above the persecuting spirit of his age altogether. The Calvinists in Holland, in England, in Massachusetts, were by the record held guilty of conscientious bloodshed for the offenses of conscience. Coming down the tide of time, what spirit ruled at Sea Grove? The hospitality, tolerance, and courtesy of the Sea Grove Christians were early put to a test, and as to the manner in which they bore the trial, the evidence of the one who gave whatever provocation may have been felt shall be admitted.

The letters here quoted from were published at Boston, Massachusetts, in the "Banner of Light," an old, ably-conducted Spiritualist journal, much respected by its supporters, and of world-wide circulation. Addressing the editor as a personal friend, the correspondent at Sea Grove freely comments as follows:

"While your various correspondents are sending you cheerful notes from different points where the thousands of liberal souls congregate and enjoy the satisfaction of a refined society and philosophical teachings, I add my scribble from another locality, where it may seem that I am out of place and ought to be uncomfortable.

"Sea Grove is a creation, and a creation by Presbyterians. If you take your United States map, and let your pen-handle run down the coast of the Atlantic southwardly until you reach Cape Henlopen, you will be fourteen miles beyond where I am. Still, Cape Henlopen light, shining across the entrance of Delaware Bay, threw its rays into my window last night, for I slept at the very end of Southern New Jersey, on the shore of the actual Cape May. The long-established watering-place of that name is north of here, in a much less desirable locality, not on a cape at all—hence, in our American way, its name.

"On the point of the Cape a few Presbyterian gentlemen and capitalists have laid out in noble style a small town, building, as their forefathers in the faith in New England did, a church first, and then, as the pilgrim fathers *did not*, a comfortable hotel, with modern improvements, next! Nor this alone, but they have leveled the sand banks, improved the shores of a small fresh-water lake, and multiplied streets, roads, avenues, and boulevards in every direction. Fine cottages have been erected, and the place is rapidly developing characteristics of material order and beauty.

"The Abraham, the Moses, the Solomon of this enterprise is Alexander Whilldin, Esq., a wealthy wool merchant of the city of Philadelphia, in whose family the land hereabouts has been a legacy for generations. Presbyterianism has descended in the same line as the property, but it must not be understood that the large fortune of Mr. W. was all inherited, or that he is of that class of men who accept their creeds ready-made from their ancestors. On the contrary, he is a thorough man of business, as liberal in his charities as thoughtfully tolerant in his adhesion to his sect. In association with him is the famous Napoleon of clothiers, John Wanamaker, of the same city. Both these persons are remarkable in the same way—men whose broad views and ceaseless energies, coupled with catholic sympathies, make mere sectarianism seem impertinent, and exalt our conception of human nature as we observe their philanthropic activity and eminent public spirit.

"The plan of this sea-side paradise, this New Jerusalem in the sand, as well as the public improvements, reflects credit upon the taste and skill of Mr. J. C. Sidney, another Philadelphian, and an architect of repute. Under his superintendence, backed by abundant means, the growth of this place has been exceedingly rapid, and yet substantial completeness is everywhere evident.

"Under the favorable laws of this State (New Jersey), the regulations of this new town are as peculiar as the old Presbyterian Blue Laws of Connecticut; in fact, they smack somewhat of their character. I should hesitate long before I consented to such laws for a State or large city, but here, and now, very possibly they are excellent; anyway, those who disapprove can go to—Cape May, or even Long Branch, which is worse.

"For my part I am glad to get to a place where rest and health and personal improvement seem really to be the object of those around me. I am rejoiced to be, even for a week, where my eyes are not offended by the emblazonry with which the rum traffic decorates so many fronts in town, and where the tippler tippleth not and the drunkard cometh not. Continual swearing (*in others*) is not essential to my happiness, while slang and obscenity, such as I often hear in some resorts, make me crawl all over with disgust. Sea Grove has no rum traffic—never will have; it has no scenes of riot, and moreover is clean and decent in every way. I don't know how rigid the regulations are, but I do know they will be enforced, whatever they may be; and that now the result is every way satisfactory if health and rest are really desired.

"It seems queer to see the people of a hotel convene twice a day for family prayers, where various clergymen 'address the throne of grace,' and a fine quartette like this of the Hayes family leads the singing; yet such is the fashion here, and I, wishing to be in style, followed the fashion. I cannot detect any demoralization in myself in consequence of this self-indulgence, and if I think my own thoughts the while prayer and music go on, I do not think I am surrounded by a company of mere hypocrites and canting, pretentious formalists! I am sure some people I know would be surprised to learn how much of real human goodness unspoiled there is in all the churches. The danger here seems to be that so much piety and propriety, 'taken straight,' may become dull from monotony, and so efforts have been made to avoid sanctimoniousness. A very distinguished Presbyterian divine organized a minstrel troupe from the kitchen and the dining room, and they gave an entertainment. Then last evening there was in the parlor an exhibition of sleight-of-hand by '*Professor Guernella* and lady.' He belongs to the assumed exposers of Spiritualism, and I have to say that he was decent in his remarks and clever in his tricks, but his *imitation* was as much like spirit phenomena as the pantomime of the deaf and dumb is like the eloquence of Wendell Phillips; and so as before the facts of Spiritualism remain, as *Guernella* says, to 'puzzle longer heads than mine.'

"To-day, after the teachings of *Guernella*, that 'we should attribute nothing to supernatural causes because we don't understand it,' we had a sermon from the Rev. Mr. Nevins upon the stilling of the sea by Jesus. He took occasion to inculcate 'muscular Christianity,' saying salvation was incomplete without health, and that Jesus healed the sick. He then told us that the storm on Galilee was the work of demoniacal spirits, as were *all* storms, earthquakes, and other destructive outbreaks of nature! It was cheap science, even if good theology; anyhow, it showed *Guernella* had not effaced the idea that somehow good or bad spirits had much to do with our life and its environment. Nevins is an elderly Presbyterian minister. To-night, at 5 o'clock, we listen to the Rev. Mr. Stockton, an Episcopalian of reform tendencies, but still in full communion. From conversation with him, I expect liberal things."

Disagreeing radically and frankly with those around him, the writer of the above states, in the further course of his correspondence, that candor and courtesy were the only concessions made by him in frequent conversations and debates with both laymen and ministers; and yet he declares that nowhere was he ever so cheerfully tolerated, never treated with greater kindness, "not even in the radical Israel."

§ Such has been the course taken by the Presbyterian managers of Sea Grove, and such is the concurrent testimony of various parties as to the order, morality, and liberal tolerance of those who frequent the place; yet there resides and rules the same Calvinism which was believed in by the Puritans, who sanctioned the death of dissenters in England and Massachusetts a few generations ago. Since then, how much of growth in grace!

Divorced from the entanglements of state ecclesiasticism, the free

Presbyterian Church, like other Christian organizations, has escaped from potent influences of corruption and gained in spiritual life. Delivered from persecution, endowed with freedom, resting secure, that Church has outgrown the old-time Puritan arrogance, intolerance, and cruelty. This is not a change in Calvinism, but it is progress among men. It was not the creed, but the fears of the Puritans, which made them exclusive and proscriptive.

The progress made manifest by the success of the Presbyterians in New Jersey has been shared by the Episcopalians of Virginia as well, and has extended to every Christian denomination in our country. It is the fruit of religious freedom, of security, prosperity, and culture; the expression of the spirit of the nineteenth century, the outgrowth of the republican institutions of the United States of America.

"Calvinism ran to seed in Massachusetts," it is said; its thorns it put forth in Europe in defiance—a defense against its persecutors. After two hundred years of tolerance and liberty, it blooms at Sea Grove; the humanities, the courtesies, the graces of life, blossom in beauty on the same rugged stock which so long has nourished the sterner virtues.

Freedom is the natural basis of civilization, progress, and a true life. Religion needs no *establishment* except in the hearts of the devout. The only *legitimate* rule is the law of equal rights, "a government of the people, by the people, and for the people"—"never to perish from the earth."

Such are our conclusions. Such the lesson of New Jersey and Sea Grove; the historic argument of SCHEYICHBI AND THE STRAND.

GEOLOGICAL OUTLINES AND ITEMS.

"A FACT IN NATURE IS AN ACT OF GOD."

"THE COURSE OF NATURE IS THE ART OF GOD."

YOUNG.

CAPE MAY LIGHTHOUSE is at the southern end of the State of New Jersey, and, according to the United States Coast Survey Reports, is in 38° 55′ 50″ .42 north latitude, and in 74° 57′ 15″ .57 west longitude; high-water mark by the same observation was 1188 feet due south of it, or in latitude 38° 55′ 39″ .65 north, and longitude 74° 57′ 15″ .57 west. The light is one hundred and sixty-seven and three-eighths miles from the northern limit of New Jersey, and between Delaware Bay and the Atlantic Ocean. Immediately west is located the settlement of Sea Grove, including the United States Signal Station at the extreme point of Cape May. Both the light and the settlement, as well as the long-famous resort of Cape May City, and the country thirty-two miles north, are included in Cape May County.

CAPE MAY LIGHTHOUSE, 1876.

Geologically, this county, in common with all the southern portion of the State, belongs to the Tertiary and recent formation of the Cenozoic period, and is characterized by deposit, drift, and alluvium. The whole county is very low, level, and uniform, and, in the absence of mines, quarries, or other deep excavations, geological examinations have been confined to the surface, and the deposit to the depth of three hundred and thirty-five feet beneath it. The best opportunities for observations have been afforded by the boring of several artesian wells at different points.

§ Not very long ago—as time is counted in geology—the ocean shore of Southern New Jersey extended from Trenton, on the Delaware, to Woodbridge, on Staten Island Sound, running nearly along the present railroad from Trenton to Metuchen, Middlesex County, and from thence eastward a short distance. This was the southern limit of the appearance of the red sandstone of the *Triassic* formation. All the land between there and Cape May, to a depth of about seven hundred and forty-two feet, in the Cretaceous formation, and one thousand or more deep the rest of the distance, has been "made" either by deposits from the sea and from vegetable growth, or by "drift" and wash of materials.

The Cretaceous formation extends from the southern line of the Triassic southwardly about sixty miles along the Delaware as far as Alloway's Creek, Salem County, and from thence northeastwardly to Shark River Inlet, on the Atlantic coast, eight miles below Long Branch. Clayton Station, on the West Jersey Railroad from Philadelphia to Cape May City and Sea Grove, is near the southern border of the Cretaceous formation.

The Tertiary formation covers all the surface of Southern New Jersey south of the line from Shark River Inlet to Alloway's Creek, except a narrow margin of recent formation along the shores. Owing to the nature of their materials, and the agencies which have operated upon them through successive ages, it is very difficult to definitely outline the field of these two formations. The Tertiary overlies the Cretaceous in the north, and runs irregularly into the recent formation along the shore.

Beyond the Cretaceous, to the north, appears the Triassic formation, composed of the red sandstones and others, the trap and conglomerate rocks.

North of the Triassic rise the mountains which stretch across the State of New Jersey in its northwestern portion. These mountains are composed of gneiss rocks and crystalline limestone, or marble, but mostly of gneiss; these are the outcrop of the metamorphic rocks of the Azoic time, and are metamorphic, igneous, or primitive in character, —that is to say, they are geologically the most ancient rocks, and owe their character to the action of fire. The valleys among these mountains are limestone localities, and all the territory of New Jersey beyond the mountains to the northwest is a limestone region, and of the Paleozoic division of geologic structure and time.

§ It being known that the metamorphic rocks of the Azoic period are primitive, igneous, or Plutonic rocks, it is understood that they are the oldest, and, if retained in place, would be the deepest buried, of all the strata, lying as an ever-shrinking shell of granite upon the fiery lava which forms the liquid pulpy heart of the globe. All the strata, were they "in place," would be piled one on another above this heated granite floor. First would come the Azoic or Metamorphic rocks; then the Silurian, Devonian, Carboniferous, Permean, Triassic, Jurassic, Cretaceous, Tertiary, Post-Tertiary; and last and uppermost of all, the Recent formations.

This would place Southern New Jersey geologically where some of its residents declare it is to be found in every respect,—"at the top of the heap;" but, since the central fire of the planet first began to cool, and islands of red-hot stone floated upon the incandescent ocean, there has been many a commotion and violent shaking-up of things in this world, and geologists are not the only people who, in order to learn the truth, are compelled to follow with painstaking care the clue of fact through what seems a labyrinth of confusion before they catch a view of the system of nature, comprehend in part the laws of the universe, and realize with reverence the glory of the Infinite God!

The geologic strata are first formed, and then upheaved or depressed, crushed, crumpled, and distorted, disintegrated, mixed, and distributed, in a thousand ways and positions, by the action of known but inconceivable forces. The shrinkage of the earth's crust crowds up enormous ridges of granite or other rocks, which in cragged wedges slowly pierce upwards through the superincumbent mass, or else the crash of earthquake explosions produces effects in a few moments otherwise the work of ages. The variations of condition and character thus introduced in the geologic elements by the causes described create an appearance of utter confusion bewildering to the uninformed and heedless. It seems to the superficial observer that the rocks and earths, the sands and the soils, are jumbled together without sequence or significance, and such persons, if induced to consider the subject at all, are inclined to surrender the use of their senses and reason, and atone for their imbecility and unfaithfulness by the acceptance of some superstitious, heathenish, and wicked pretension of a revealed Cosmogony. Thus they give up the study, appalled by the difficulties which surround it, content to know no more scientifically of the wondrous world they mysteriously inhabit than did the saurian reptiles whose fossil remains enrich the marl beds and banks of fossil shells. Such a course is blindly impious, and disgraceful to human nature.

It is true the Bible asserts that God made the world, but it gives only the most exceeding vague intimation as to how or when the Creation announced was effected; whatever may or may not be revealed in spiritual things, we are left to study geology hammer in hand, knocking hard at the rocky doors of science. Yet we need not be discouraged nor afraid; the difficult is not of necessity impossible, and although geology is an infant science compared to astronomy and mathematics, and only a child beside even chemistry, yet the clue has been discovered, the system made plain, and only diligence and courage are required for the conquests of the future. The practical economic value of geology is immense, and besides, it must ever be a high gratification to read in the record of the rocks the history of the evolution and progressive development of our home, the earth.

There is no danger the facts of Geology can annul, or even obscure, the truth of religion. Men of science are not always scientific, but while we trace the process by which that which is has been brought about, it need not be that we become process mad, and unable to see in and behind the unfolding the INFINITE SPIRIT, which moves in the wheels of existence. Here, are phenomena; there, is law, process, and evolution; THE SPIRIT is everywhere, all in all. There is no pebble so small but *law* constrains it, no material so inert but evolution compels its progress; the smallest grain of sand has being in an infinite order; omnipotence overtops the loftiest crag, underlies the deepest primitive strata, and sustains the central fire. Facts cannot disprove TRUTH; the idea of God science can displace from the minds of candid and thorough students is but the myth of morbid imagination, the shadow of the fetich of barbarian ignorance.

§ New Jersey contains the out-crops of all the geologic formations except, unfortunately, the carboniferous. In the absence of the coal-bearing strata there are, however, other rich and rare mines and deposits in the State, notably those of zinc; as well as an abundance of iron, of lime, of valuable clays, of building materials, and natural fertilizers. The remarkable geologic characteristics which have marked the region, are the evident recurrent upliftings and subsidences of a large part of the surface, and the effects of denudation and drift.

The Cretaceous formation of New Jersey, which with the Tertiary covers the whole southern part of the State, was once the bottom of a shallow and quiet ocean; it is evident from its stratification that the surface of the land rose and fell with comparative regularity, so that the sea would advance at times and cover it, and then the bottom would be uplifted and the sea recede. Vegetation would start up upon the marshes and upland, which would after a time by subsidence of the land be overwhelmed in the waves, and then buried by degrees in the sea sediment.

In proof of all this, the immense quantities of fossil shells in this formation are found unbroken, and the bones of reptiles lying together undisturbed near where they lived and died; this would not be the case if the sea which covered them had been turbulent and stormy. That marine shells and sea sediments are found both above and below various beds or layers of vegetable fossils and the bones of land reptiles shows that alternately the land was submerged, and then for an age emerged from the waters. Yet all this time the land must have been subsiding on the whole; for a regular stratified formation some eight hundred feet thick was thus aggregated, and the topmost layer of shells was of course under the tide when it grew. This subsidence was followed by an elevation of the whole coast to about four hundred feet above the level of the sea, which was effected bodily; but the uplift seems to have been greatest in the northwest, so that the strata slope or "dip" toward the southeast at present. This upheaval was before the "drift period." When it came, the process of denudation reduced the land to nearly its present level and configuration.

The name of the Cretaceous formation is derived from England, and is significant of the great amount of chalk which characterizes it. The constituents of the formation in New Jersey are all earthy, except where in a few detached spots the material has become cemented by oxide of iron into a kind of sandstone or conglomerate. The strata are the upper marl bed, the yellow sand; the middle marl bed, the red sand; the lower marl bed, clay marls, and plastic clay. These last are of fresh-water origin, and are supposed to have originated from the decomposition of gneiss rock; they are the underlying strata when in place, but in actual situation crop out on the surface on the northern edge of the formation at Woodbridge, Perth Amboy, South Amboy, Washington, and Trenton; they are also used as potters' clay at several other places.

The other strata came to the surface one after another as distance increases towards the south, until at the commencement of the Tertiary formation the upper marl beds appear while the other strata are mostly subterranean.

Next to the evidences of denudation and drift presented by the surface of the Cretaceous district, the vast quantities of fossil shells and bones are remarkable. The shells of the clay beds are of fresh-water origin (such as the genus *Unio*, as fresh-water mussel and others), and may have grown at the bottom of lakes before the subsidence, or the fresh water may have been kept from the sea by hills and ridges. The green sand which abounds in the Cretaceous formation is supposed to have become granulated by forming inside very small shells, and is of chemical origin, and evidently a deposit from salt water, as the vast amount of fossil marine shells contained in it demonstrates.

One species of these shells, the *Terebatula Harlani*, forms a layer ninety miles long, over a mile wide, and about a yard in thickness in the middle marl bed. This layer is made up almost entirely of this species of shell, closely packed together. Immediately beneath the *Terebatula Harlani* shell layer is another equally large, made up of shells of the *Pycnodonta convexa*. Many other kinds of shells exist in great quantities in the Cretaceous formation at various places. Of these over three hundred varieties have been classified and described, with no certainty that the work is complete. In some marl beds a dozen or twenty varieties might be found in comparatively small space, and then again, as before described, beds of one kind of shells, a mile wide and several feet thick, are scores of miles in length.

The plastic clays of the Cretaceous formation of New Jersey are highly valuable to the potter and to the maker of fire-brick; the common clays are useful too for ordinary brick-making; the gneiss of the Azoic formation, and the red sandstone or "brown stone" of the Triassic strata, and, very generally, the various colored limestones of the Paleozoic district of the State, are used in building; the brown sandstone of southern New Jersey serves the same purpose. The Triassic trap rock and sandstone is used in paving, and its slates for roofing. Iron and zinc mines are very rich in New Jersey, the iron produced being of the best; the zinc ore is generally rare elsewhere, yet ten years ago twenty-five thousand tons of it were dug yearly in Sussex County. This ore yielded seven thousand tons of "zinc white," and five hundred tons of metallic zinc; this was seven-tenths of all the zinc white manufactured in the United States, and about one-fourth of all the spelter produced. Yet it is probable the marl beds of the Cretaceous formation, used alone as fertilizers, or in combination with its shell and stone lime, and the muck and peats of the region, are or might be made worth more than all the quarries and mines within the Commonwealth.

The green sand marl was first used as a fertilizer, in Monmouth County, in 1768, when "an Irishman" ditching for Peter Schenck "threw out a substance he called '*marl.*'" It was spread over an acre and a half of land, where its good effects were visible for many years; "but," says the record, "this circumstance attracted no particular notice until 1811, when the farm came into the possession of John H. Smock;" then notice was taken of the effect of the marl, and the use of it began in the neighborhood. It had been used somewhat at that time in other places, but at no place in this country was the use of marl general before the present century began.

The discovery and use of the marl have raised thousands of acres of lands from sheer barrenness to remarkable fertility; worn-out farms, where a family could not be supported, are now making their cultivators rich by their productiveness. Bare sands are made to grow clover, and then crops of corn, potatoes, and wheat. "Pine barrens," by the use of marl, have been made into fruitful lands, and thus whole districts have been saved from depopulation, and the inhabitants of others increased.

Fifty-five years ago the six southern counties of New Jersey were described by Morse as four-fifths waste and barren land; this constituted two-fifths of the entire State: now, large portions of this desert are under high and profitable culture, and the land in farms in the six southern counties is worth an average of over fifty dollars an acre.

The Irishman who spread the first marl in New Jersey deserved more honor than many a conquering warrior; a monument erected to his memory would be more in keeping than to have him referred to in the State Geologist's Report merely as "an Irishman!"

In the marl-beds of the Cretaceous formation are abundant and extraordinary remains of extinct reptiles. They were of the orders THECODONTIA, SAUROPTERYGIA, TESTUDINATA, CROCODILIA, and DINOSAURIA. A fine specimen of the last order is preserved in the Museum of the Academy of Natural Sciences at Philadelphia, and is classified as the *Hadrosaurus Foulkii.* It was a gigantic reptile, about twenty-eight feet long. The hind legs were very long, more than double the length of the fore limbs. The reptile walked on his hind legs as a man uses his legs, and ate foliage and vegetable food. It was a heavy unwieldy monster, living on land or in the marshes.

There were carnivorous reptiles also, some of them forty feet long, some *only* about twenty-five feet long, with a body as large round as an ox, and a long neck. These steered themselves by flippers like those of a whale, and propelled themselves by their tails. Some of them had flattened tails, and *sculled* themselves along as a boatman uses a single oar; some of them had great conical teeth: they ate fish probably; such were the Cimoliasaurus, the Elasmosaurus, the Mosasaurus, and the Clidastes; the last, however, was more serpent-like, and fifteen feet long.

There have been more than twenty kinds of Tortoises, Turtles, or Terrapin found. One of them, the *Euclastes,* was full six feet long, and very strongly constructed. Others were as large, and some had exceedingly thick shells, notably the *Adocus Petrosus* and *Adocus Firmus.* The Crocodiles, Alligators, and Gavials were very numerous; three-fourths of the bones found are of this order, and the wonder is what such swarms of them lived on, as they have left no remains of their feasts to tell the story so far as yet seen. These horrid brutes were twenty feet long in some cases, but they varied in size, some being four feet long only.

The *Dinosauria,* of which order is the Hadrosaurus Foulkii, were the highest order of reptiles, and in some characteristics resembled birds. Many of them were as large as Mastodons and Elephants. Some of them squatted; some jumped like the Kangaroo; some, with great long legs, stalked around flopping their half-useless arms, and overlooking the levels with bird-like eyes set in a "bony visage," as if their face was trying to become a beak!

Such were the monsters of the Cretaceous land, such the shells once alive, when it was under the sea. These reptiles, and the vegetable remains in conjunction with them, indicate a torrid climate; but the bones of *Walrus* have been found in the neighborhood of Long Branch, and it is otherwise evident that their long summer was, in the "drift period," turned, and perhaps suddenly, too, into an equally long and appalling winter.

§ The phenomenal effects of Denudation and Drift are not confined to any geologic formation or geographic locality, but may be observed throughout many extensive sections, and indicate the force of several agents acting at separate times upon diversified materials in different directions, and by various modes and in distinct degrees. These agents are evidently three,—wind, water, and ice,—and the mode of action by each is unlike. Exposure to the heat of the sun and the effects of cold and frost disintegrates the rocks and subjects them to the effect of drift-creating forces. The common variations of climate and season are efficient in this respect, but special causes in different ages have vastly intensified the influences of temperature and weather.

The influence of wind and of water is constant; but the vast effects of the ice-drift are referable to the geologic "Drift Period." The force of wind is active not alone in wearing away rocks, by whirling grit and transporting great quantities of sand, and building dunes and beaches along low shores, but in some places it wafts the sands of shores and deserts far over fertile fields and even forest hills, thus sadly increasing the area of sterility. In the African deserts the awful simoom blights vegetable growth and suffocates animals and men, then lets fall over the dead caravan thick layers and hills of sand for their winding-sheet and grave.

On the Western American plains, and among the mountains of that region, the winds have cut countless cavities into solid stone; these cuttings vary from small orifices and hollows to large channels and openings; in fact, in some localities the most of the strata has been worn away, and only small isolated elevations of fantastic form remain to denote the former level of the surrounding territory. On loose sand the operation of wind is obvious: the finer earth and dust is lifted and bodily conveyed to a distance in proportion to the strength of the blast. while the coarser sand and gravel is rolled, slid, and *drifted* along the surface, often up steep inclines and considerable elevations.

As a gale grows in violence, the power of wind increases in the same degree to an unknown limit: typhoons and cyclones exhibit its force in Indian seas, the West Indies are often devastated by hurricanes, and in portions of the United States whirlwinds and tornadoes sometimes level giant forests in their path, demolish strong buildings, and hurl the ruins far through the air. A hurricane in the West Indies broke down a very heavy wall, and rolled stones weighing hundreds of pounds along the ground!

By forcing a blast of air through a nozzle, and charging it with sand, made to impinge upon flint glass, artisans abrade, cut, grind, and engrave the glass most rapidly. In a similar way, the wind, forcing itself through rocky canyons, notches, passes, defiles, fissures, and crevices among the mountains, and sweeping over rainless plains, takes up the gritty débris and sharp sand and, *whirling* them along, drives them in an enormous rotating *sand-blast* against the rocks. Gneiss and adamant could not resist the impact and continued friction.

The effects of water and ice in the Drift Period have been closely studied and elaborately stated by the geologists, but it is possible the effects of winds have not been as fully observed and noted. By some commentators it is supposed that the destruction of the Assyrians (II. Kings, 35) was accomplished through the agency of a simoom, which certainly would be a sufficient natural cause for the death of even that host. However it may have been in this case, there is evidence throughout the scriptures of the Old and New Testament, but not confined to the Bible or any book, of the power of spiritual beings who, under Providence, use the forces of nature as their instruments to do the will of Heaven. No marvelous work is belittled or made less wonderful because we are enabled to discern the agencies and the method by and through which the Eternal Power is made manifest.

The evidence of the former submergence of vast areas which are now the elevated portions of continents, and of tremendous floods which have deluged the surface since it has been uplifted, appears almost everywhere, and seems to be amply convincing: tradition among savages, and the poems and mythological records of many races, refer to such phenomena, ascribing them generally to the action of the gods. The Bible account of the deluge, it is thought by some, finds corroboration in these legends and poetical allegories of antiquity. Certain it is that water, in showers, floods, and oceans, has been the potent cause of distribution and change in geologic materials. When porous stones are exposed to rain and severe frost, they rapidly disorganize: the water penetrates the pores of the stone and is frozen there, the expansion of water changing to ice bursts the cells of the stone by the exertion of one of the most potent natural forces, and the rock soon crumbles from the effect of such weathering. Some rocks, when submerged or long subjected to the action of water, become "rotten stone," the cementing material in their composition being oxidized or dissolved away. In turbulent torrents the stones are dashed against each other and broken, they are ground together and pulverized, and, after trituration, are borne away to form the sediment of quieter waters. Thus the winds, the rains, the streams, and the waves co-operate, and through their action, in time, the rocky mountains are reduced to a bed of sand, to be drifted about by every flood or borne away before the wind.

The influence of changing weather and seasons is incessant: every warm day, every wandering wind, every passing shower, is active in changing the surface of the earth, while geologic indications prove that not only have icy oceans rolled over what are now the mountain-tops of temperate climes, but glacier-like formations of ice, during the winter ages, crept down from the pole, submerging the life of the zone beneath a curtain of frozen death. Gripping immense boulders of flinty rock in their icy flow, the frigid seas dragged them for hundreds of miles, firm fixed in the icebergs as a glazier's diamond in steel. Grinding heavily on the bottom of shallow seas, these enormous tools in the hands of Nature have scratched and scored the granite mountains, the trap dikes, and the various ridges, until they have, in some cases, been utterly worn away under the long-continued and terrific abrasion and their débris scattered far and wide. From astronomical causes divergences are supposed to occur in the polarities of the earth, producing excessive and sudden but persistent changes in climate, or, as is known, comparatively slight deflections of constant winds and currents gradually bring about the same result. In the far north mastodons by thousands are to be found imbedded in ice, where and as they stood when the torrid climate congenial to them passed away at once, and paralyzing frost and overwhelming snow descending upon them established most abruptly the conditions of Arctic winter.

§ The Drift of New Jersey indicates not only a grand movement of the agencies of denudation from the northwest, but counter, or rather divergent, currents of a similar nature, due to local elevations or other secondary causes; the main line of advance however being toward the southeast.

The Paleozoic formation to the northwest of the mountains was of course the first affected by the southward tending drift. The amount of material displaced is almost incredible. One body of drift in the Paleozoic district is one hundred miles or more long, from ten to fifty miles wide, and two to three thousand feet in depth.

The drift action in the Azoic formation has also been immense, and the evidence of it is to be seen not only among the gneiss-crowned mountains, but all over the State, as the disintegrated granite appears everywhere in almost every foot of gravel bed. On the lower margin or southern border of the Triassic formation a belt of gneiss rock is exposed; this was drift from Azoic outcrops in the mountains, and has aggregated in its present place and concreted into stone, and then again has been in part abraded, disintegrated, and carried away. The most common soil of the Azoic formation is drift, deposited among the gneiss rocks and mountain ranges. There are limestone boulders in the neighborhood of the gneiss which weigh two thousand tons each, and which have drifted a mile at least, and perhaps several miles, and have been *lifted* one or two hundred feet. On Sparta mountain, twelve hundred feet above the sea, are found boulders which weigh a hundred tons, which have been carried there from an unknown distance. Boulders of ore have been carried into distant deposits far from the original strata, and have misled those who found them into the idea that they were indications of mines in the place where they were discovered.

It is in the Triassic formation, however, that the greatest signs of drift action appear; there the red sandstone has in many places been worn away from two to five hundred feet. The Newark marshes have been dug out by drift action, and the excavation was carried below the level of the sea, resulting in bays now but partly filled by mud, grass, roots, etc., etc. Boulders of various kinds appear in this formation. There is a boulder of five hundred tons' weight on the northwest slope of First Mountain, near the Newark and Mount Pleasant turnpike, which has been carried by the drift current full thirteen miles. There is another boulder near Woodbridge, more than twenty miles from its parent strata, which must weigh two hundred and fifty tons. There are trap and sandstone boulders everywhere in the Triassic formation, and considerable deposits of limestone in loose masses. Paleozoic fossils are also found scattered with the drift into Triassic beds.

The Cretaceous formation has been worn away and changed by denudation nearly as much as the Triassic strata; Naversink Highlands, and the Mount Pleasant Hills of Monmouth County, have perfect seashore pebbles upon their summits; yet they are about four hundred feet high, their valleys being one hundred feet above the sea-level. In the Cretaceous beds and layers, the forces of the drift encountered only friable materials, as the stratification was comparatively recent, and the elevation referred to in a former page still later; hence, when these forces became active among the shell layers and the loose sands which had been imported from other formations and not concreted, the uniform surface of the uplifted land was worn into valleys, or washed away entirely for large areas, to the depth of three or four hundred feet.

Thus were created the broad low plains now visible in South Jersey, and the low hills with their shallow valleys between them, which there mark the Cretaceous area. The sides of these hills, and the bottoms of the valleys, afford excellent opportunities for geologic observation. The sand, loam, and general mass of material dislodged, was carried away toward the south. The drift action continued a long time, and thus were the abraded constituents of all the strata to the north mixed with the materials of the Cretaceous sediment, swept out to sea and deposited there, creating the Tertiary formation.

That extensive areas should rise and fall, and even "the sit-fast and immovable hills" appear and disappear, grow and waste away, seems incredible to the untaught, and is wonderful to all; and yet in geologic ages continents emerge from the sea and then sink again beneath the ocean; Himalayas, Alps, Andes, and Sierras swell aloft by the action of geologic forces, and then subside into the subterranean, or are sculptured into picturesque forms and worn away by *denudation*.

Geology gives time; and in time, the sun, the rain, the wind, and the frost, as has been demonstrated, will humble the head of the highest mountain that lifts its granite top above the clouds! Then again by *sudden* or by gradual shiftings of the balance and polarities of the earth, tremendous and sometimes abrupt changes of climate have been induced, and vast floods and moving fields of ice have been the consequence. These awful forces, *in time*, work out the grandest results and most radical changes.

The whole solid crust of the earth, moreover, is no thicker in comparison to its liquid, fiery mass than the shell of an egg to its contents; therefore any changes of the surface are not unnatural, or, in view of the facts, a matter of amazement. The perpetual miracle and admirable wonder is that, with such forces always in action in some form, the universal equipoise is maintained, and the conditions of human life and happiness evolved, with Infinite wisdom, from the perturbations of nature!

§ Formerly, the names of Primitive, Transition, Secondary, and Tertiary, were applied to various kinds of rocks, in geological classification; modern usage substitutes the technical terms of Azoic, Paleozoic, Mesozoic, and Cenozoic, to define periods marked by peculiar stratifications and fossils. The rocks themselves are now called Metamorphic, Silurian, Devonian, Carboniferous, Permian, Triassic, Jurassic, Cretaceous, and Tertiary; the last significator being retained and adopted from the old terminology; besides, the phrase *Post Tertiary* is used, meaning since the *Tertiary*.

The rocks are named with regard to their constituents and character, or derive their titles from geographic localities where they especially abound, or where they were first scientifically observed. The Azoic rocks are supposed to have been formed before vegetable or animal life existed on this planet, and until the last few years it was supposed that no form of life was known below the lower Silurian rocks, that is to say, in Azoic time, before the Paleozoic period. American geologists are disposed to admit that the recently investigated "eozoön" found below the Silurian is the fossil of an animal form; if they are right, the history of life on earth must be antedated, and carried back very far in time, and an Eozoönic age be recognized between the Azoic and Silurian.

The Paleozoic period was the age of Mollusks or shell-fish, and their fossil remains abound in the limestones of the Silurian division. The Devonian sandstones and shales contain shells and the fossil remains of vertebrate fishes. The Carboniferous division of the Paleozoic time has no place in the geology of New Jersey, it is exceedingly developed in Pennsylvanian coal measures; in its time land plants flourished beyond comparison; these fossil plants are coal at present. In the Triassic, the Jurassic, and Cretaceous divisions of the Mesozoic time there was an enormous development of Reptilian life, and the bones of monster reptiles are plentifully found as fossils in the rocks of those layers. The Mammals, which are warm-blooded quadrupeds, appear as fossils first in the Tertiary rocks and strata, and the human period, the time in which man has inhabited the earth, is included in the *Post Tertiary.*

The divisions of the Tertiary formation are the Eocene, the Miocene, the Pliocene, and the Post Pliocene, including the recent; in New Jersey it constitutes the formation of the territory south of the Cretaceous strata, being bounded on the north by an irregular line from Shark River to Alloway's Creek. The Recent formation lies along the sea-shore and the banks of various streams, and generally includes all lands less than twelve feet above the level of the tide.

None of the boundaries of the Tertiary fields are sharply defined like those of the rocky strata; the drift and wash has intermixed the materials of the formation, merging the outlines of the various beds and layers. Though the earthy nature of the Tertiary formation subjects its surface to change from storms and streams, by which the beds are mixed together or discolored, yet the mineral substances therein are undisturbed in their original places of deposit and not petrified, while even the lowest Tertiary strata contain fossils of existing species, proving the modern origin of the whole. The upper marl bed is in the Eocene division of the Tertiary formation, and is the lowest layer of its stratification.

In a well bored at Winslow, Camden County, New Jersey, there was found:

First 5 feet of surface earth.
Then— 15 feet blue and black clay.
95 " glass sand.
35 " miocene clay.
107 " micaceous sand.
43 " brown clay.
A gum log one foot thick. (!)
20 feet green sand, marl, white shells, teeth, etc.
15 " pure green sand.

At which point water rose from the bottom of the green sand. This gives a good general idea of the structure of the Tertiary formation in New Jersey.

Loamy clay, white quartz pebbles, silicified fossils, feldspathic rock, etc., intermixed with sand, the materials of the drift, overlie the other beds unless the surface has been washed away. This drift varies much in constituency from pure clay to clean sand; it is generally reddish yellow from oxide of iron, often fertile and retentive as a soil, and makes good roads, packing into a solid, smooth, durable bed, even when spread over loose sand. The excellence of the road-bed of the avenues of Sea Grove is due to the liberal use of this material upon them.

The glass sand underlies the drift gravel to the depth of ninety-five feet, and is pure white quartzose sand, except when it comes to the surface or at the bottom of the bed in places, then it is sometimes discolored. Its use in glass-making is very common and important, and this section, from which much glass sand is now shipped, contains enough of this valuable material to supply the world for a thousand ages! The sandy plains of South Jersey are the exposures of this bed of sand where the drift gravel has been washed away.

The Miocene clays and marls of South Jersey, so largely and successfully dug as fertilizers, contain numerous fossils, and are a source of wealth as well as a matter of geologic interest. The Micaceous sand, one hundred and seven[illegible] feet deep, found in the well at Winslow, does not crop out at the surface anywhere, and is in place below the sea level. The same is true of the brown clay found as described.

Thus the layers of the Tertiary were formed, being deposited as drift from the more ancient strata. Mixed with the Tertiary layers, or distributed through them, may be found constituents of all the older formations in the State; thus thrown together, they have, by chemical action and reaction upon one another, entered into new combinations and produced new substances; these in turn, with all the rest, subjected for thousands of years to the play of elementary forces, have been variously manipulated and chemicalized continually, while all the time impelled by the floods and streams toward the sea, along whose shallow margin they have been deposited, forming in "Recent" ages still another new shore to the ceaseless waves.

§ As the Tertiary formation is marked by drift and earthy deposit, so alluvium characterizes its Recent division. In the Tertiary we find clay, sand, gravel, loose pebbles, and some boulders, most of which are from distant strata; the beds of the Recent are made up of finer sands and clays, loams, mud, peat, etc., derived from adjacent deposits or the remains of vegetable production. The fossils of the Recent are all identical with existing species; among them human remains and relics are frequent. The Recent formation in New Jersey borders the Atlantic from Sandy Hook to Cape May, and forms the shore of Delaware Bay up to Salem, also the banks of some of the rivers and creeks. The sand beaches, the marshes, the cedar swamps, and an indefinite amount of upland border in the State are recognized as being included in this formation, and are in process of formation and change. The general surface soil of the upland border is a fine sandy loam with but little gravel, and contains organic matter enough to render it productive and fertile ground. An example of such border land is to be seen adjoining Sea Grove, and forms the Stites farm. The farm has been for some time occupied by the Hon. Downes Edmunds, and has been worked in places constantly and successfully for a hundred or more years without any manure or dressing whatever, and yet has not been at all impoverished. The land thus cultivated is so full of shells in spots as to make ploughing difficult; the sub-soil is a deep, black, sandy mould.

The tide marshes of the Recent formation of New Jersey are a remarkable feature; there are about three hundred thousand acres of such marshes in the State, and Cape May County alone, with a total area of one hundred and seventy thousand one hundred and seventy-one acres, has fifty-eight thousand eight hundred and twenty-four acres of tide marsh, including ten thousand four hundred and forty-three acres of sounds, bays, inlets, etc. The marshes are but little above ordinary tide level, and covered with grass, reeds, and coarse sedge, but treeless. Beneath the surface of the marsh there is from a yard to forty feet of mud or soft earth, with an average depth of twenty feet. The marsh is deepest back from the beach and from the banks of streams, water-courses, etc. The body of the marsh is merely a bed of fibrous roots; near the beach, sand is intermixed with the roots, and along the streams and water-courses mud has been deposited, and is retained among them.

The marshes enlarge by encroachment in places upon the wooded upland, and by growing into the sounds and waters they enclose; at the same time the sea and bay have during the last century cut away many acres of the marshes, which have become exposed to the waves by the demolition and shifting of the sand dunes and beaches. The surface of the marsh, when enclosed by beaches, or by the clayey banks of streams, sinks slowly, by the decay and compression of the fibrous mass of which it is mostly composed. The wash of streams and the drift of the sea sand landward tends to solidify the marsh, as vegetable growth and deposit elevates the surface; however, shutting the water off from a true marsh causes it to sink, as it is really afloat, in and through the water, and it is so unsubstantial that many cubic feet of it when burned make but a very small quantity of ashes; the marsh is, in fact, a sound or cove choked full of fibrous roots and vegetable deposit. Many hundreds of acres of that which was cedar swamp is now salt or tide marsh; the trees having been killed by the encroachment of the sea water, have fallen, and are now buried, but undecayed, in the deep mud, the surface growth flourishing evenly above them.

§ In treating of the Cretaceous formation, on a former page, it was stated that alternate elevations and depressions of the shore line had taken place, until finally, before the drift period, the surface of the whole formation was lifted several hundred feet above the sea, from which it has been degraded by denudation and drift down to its present level and configuration. It can be readily and definitely shown that similar but less extensive fluctuations have taken place in the Tertiary and Recent formations and are now operative along the present shores. How far inland the action may reach, or in what degree affect the interior, is more difficult to decide.

In various elevated positions in the Recent formation marine shells of the common species, or casts of them, are to be seen in their natural attitude; on the banks of Maurice River, at Tuckahoe, and elsewhere along shore, they lie from eight to twelve feet above high-water mark, and indicate an elevation before the depression now going on; and as the amount of subsidence at present is about seventeen feet on an average, as estimated by measurement from tide level to the lowest points where buried and submerged trees are found in the places in which they grew, the former elevation must have raised the surface from twenty-five to thirty feet.

The highest land of Cape May County is but about forty feet above the level of the sea, and that only at a few points of very limited extent, the average elevation being but eleven feet; so that when the shells now from eight to twelve feet above tide mark were at the level in which they grew, the greater part of Cape May County must have been submerged. The last elevation carried the shore line at least seventeen feet above where it now is.

The operations of the Sea Grove Association in clearing and grading the remarkable sea-side resort they have so well begun, have obliterated some of the most interesting and characteristic traces of geologic action to be found in the State. Between the gateway of Sea Grove and the bay, and lying along the shore, were formerly a number of well-defined parallel ridges of drift sand. These were the evidence of a former uprising of the shore, and of the consequent receding of the water of the sea, which must have washed the gravel bank or fast land; the ridges were created, one behind another, by the wind, which, blowing across the ancient strand, would raise the innermost ridge first, and then, as the shore widened toward the sea, another between it and the water, and so on.

The beach ridges, having been formed long since, were covered with a heavy growth of black oak timber, which has been in part removed by the Sea Grove improvements; the parallel ridges have ceased to advance seaward, but Mr. Alexander Whilldin, a close observer, affirms that at present Cape May Point is growing out into Delaware Bay, by the deposition of sand upon it from the ocean front, and by the action of the wind piling up dunes or sand-hillocks.

Almost entirely along the shore of New Jersey, the main or "fast land" is separated from the sea by salt marshes of three miles or less in width; outside of these, next the sea, occurs a row of long, narrow, somewhat elevated, and more or less wooded islands, or "beaches." These are the Old beaches; they are more ancient than the marsh, and are supposed to have been formed during a former period of depression. The waves beating upon a friable shore of earth and sand, such as then existed, would wear a channel next the shore, and pile up a shoal outside the surf; a series of such shoals would thus be formed parallel to one another, and when the next elevation occurred they would appear above water, and form the basis of the present beaches. Shrubs and trees would soon grow upon these ridges, saving them from drifting away, and causing them to retain all the sand the wind blew from the strand upon them. The lower grounds between the ridges would finally rise above water, and presently become covered with vegetation, until a subsequent depression again carried them below tide level, when they would become salt marshes, filling with mud by the action of the tides, and keeping their surface at high-water mark, by the growth of peat, just as one hundred and seventy-eight thousand two hundred and forty-wo acres of such formation, lying all the way from Long Branch southward along the shore to Cape May, are now doing.

Fresh marshes form in the broad shallow valleys of the slow-moving rivers and creeks of South Jersey; as at Tuckahoe, and on Great Egg Harbor Rivers. The salt marshes on Delaware Bay shore have been formed as fresh marshes in the valleys of the streams which flow through them to the bay, and are supposed to have had beaches between them and the bay, which have gradually been washed away; in the same way a large portion of the marsh itself has gone.

§ The landward beaches which join the marsh are developed in long, parallel lines, and, where the timber has not been removed, are covered with a very old growth of it; the open spaces in the depressions between the beaches are called *savannas;* in wet seasons they are saturated or more or less covered and filled with fresh water; they are then called *slashes*, and are the haunts of numerous water-fowl and game-birds, which makes them favorite resorts of discriminating sportsmen.

The Old beach ridges are not over a rod in width, and not more than five or six feet high; yet they, with the savannas beside them, may be a mile or two long. The Old beaches contain a small portion of clay with their sand, which partly saves them from drifting with the wind, and promotes the growth of the timber. The Old beach varies in height, increasing in elevation toward the sea; part of the low landward ridges have become submerged, and yet can be traced in places by the lines of dead trees standing in the marsh.

At Sea Grove the marsh disappears from the Delaware Bay front, and the Old beach has formed back directly over the marsh or against and upon the upland. Lily Pond, or *Lake Lily* as it is now called, occupies the place of what might be a marsh, and yet is a fresh-water pond, from which water was formerly taken for shipping.

The water of Lake Lily had connection with the sea by a water-course which ran from the shoreward end of the lake, between the strand and the lighthouse, and along the foot of the upland, to the west of Cape Island, and so into Skillinger's Creek and under the bridge to Cape Island

Sound, and out by Cold Spring Inlet to the Atlantic. Now the sand drift has filled and covered the water-course near the lighthouse, and the Sea Grove Association have obliterated the natural features of the lake. In their zeal for improvement they have detracted very much from the scientific interest and value of the original pond, which is less to be regretted however, as there are enough indications all around of the same purport, and the engineers, by grading the shore of the pond to a fine drive, putting up ornate boat-houses, etc., have succeeded in making a very pretty miniature lake of what was a somewhat unsightly even if pure and interesting sheet of fresh water.

Since its improvement as above stated, Lake Lily has become one of the attractions of attractive Sea Grove, and is a great addition to the pleasure of visitors. When the water-course referred to was open, it was not very uncommon for the sea, in storms, to throw its waters across the beach into it, making the waters of the lake brackish; but now the natural and artificial filling in of the southern end of the water-course and the lake prevents such an occurrence, so the lake has been carefully cleaned out, and stocked with valuable fish. The waters of Lake Lily are solely from the rainfall; they percolate slowly down and out from the bed of the lake, displacing the salt water which infiltrates the sand, yet not mixing much with it. Similar effects are produced among all the beaches. The different gravity of salt and fresh water has an influence upon the phenomena; the fresh water being lightest, remains at the surface, and can be obtained by digging a few inches beneath the sand, anywhere between the beach ridges. Lake Lily is the only similar body of water on the Cape below Cold Spring.

§ The Recent formation of New Jersey, especially in the southern part of the State, is noted for extensive swamps and marshes. Those of the interior are heavily wooded, but none of them are much above tide level; the more elevated and solid are "timber swamps," and not only furnish good and desirable lumber, but might in many cases be improved by clearing and culture, and thus make valuable farms. It seems remarkable more has not been done for the agricultural development of the interior of South Jersey, but the original settlers looked to the sea for their highway, and to a great degree for their harvest too; for which reason they made their homes along the upland of the shore.

Of late, through the enterprise of several parties, notably that of Charles K. Landis, of Vineland, the interior of the State has been better appreciated, and, being extensively and judiciously advertised, has attracted many intelligent and industrious settlers, who have successfully planted many fine vineyards, orchards, and farms.

The cedar swamps, which are extensive on the banks of the rivers and around their sources, are overflowed, not stable land like the timber swamps; the White Cedar (the *Cupressus thuyoides* of the botanical nomenclature), which holds exclusive possession of them, flourishes only in submerged or saturated soils. In many places in South Jersey it grows in a peaty stratum, where there is neither clay, gravel, loam or mud, but only a compact mass of fibrous roots, and the débris of its own fallen growth. In such localities, as well as where more substantial components partly form a true soil, the white cedar grows densely, and in its young growth rapidly; afterwards it becomes crowded, and grows tall, but increases more slowly in diameter.

The vegetable remains which fall from the swamp trees into the wet mass are shaded from the sun by the evergreen foliage, and thus kept cool and saved from rapid decomposition. Settling gradually down, they become submerged and then buried, from which time their decay is almost imperceptible. In this way the surface of the swamp is gradually elevated; a layer of more than a foot thick has thus been formed in sixty years.

The original growth of cedars were sometimes seven feet or more in diameter, and immensely high; the average size of the full-grown trees, however, was but about two feet and six inches. There are none of these great trees left, and as the whole area of Cedar Swamp is cut over every second generation, or every sixty years, a living cedar tree a hundred years old is now a rare specimen; still, the natural term of the tree is a lifetime of successive centuries. Various parties have counted the annual rings in the logs and stumps of cedars, and various witnesses affirm the existence of from five hundred to over a thousand of them in a single specimen. Sir Charles Lyell, F.R.S., quoting a newspaper article of Dr. Beesley, of Dennisville, says (Second Visit to United States, vol. i. page 34) that "Dr. Beesley, of Dennis Creek counted 1080 rings of annual growth, between the centre and outside of a large stump six feet in diameter;" this grew atop of a *previously fallen tree*, which was half as old; thus fifteen centuries were registered in a couple of logs on the surface of a swamp, which has been sounded in places from *eight* to ten or even eleven or more feet deep, and is *full* of fallen logs to the very bottom.

The white cedar, though a very tall, slim tree, sends no roots down into the firm soil underneath the swamp, but spreads them laterally in the shallow, soft, black, peaty, wet earth which is its congenial place of growth. The timber standing in a natural ancient cedar swamp is but a fraction of the quantity which has fallen and become subterranean. The living timber thus buried is apparently indestructible, and has been *mined* from its place of deposit buoyant and sound, and used for the best quality of lumber, many hundreds and perhaps thousands of years after it had grown. This mining of timber has been carried on as a regular business in the swamps about Dennisville; between nine and ten thousand dollars' worth of shingles, at fifteen dollars a thousand, have been manufactured in a year from logs thus exhumed. The production of shingles did not consume all the timber taken, as a part of it was large, fine logs, more valuable for boards, into which it was sawn. More than forty thousand dollars' worth of cedar rails and lumber are produced by these cedar swamps every year, and an acre of good swamp, fifty years in growth, is worth from five hundred to a thousand dollars. The cedars are mined not alone in the growing swamps, but in meadows where only stumps and dead roots break the surface, and in places where a smooth turf entirely hides all traces of wood from surface observation, as well in a part of the tide marshes, which were once cedar swamps, but where the growth of timber has been stopped by the encroachments of salt water in consequence of the subsidence of the swamps along the shore. Of course many of the buried trees are unfit for use. Those which grew when the swamp was shallow and the roots of the trees touched the gravel bottom, are so *gnarly* as to be unfit for splitting. Some of the trees fell only from extreme age, deadness, and partial decay: these are worthless; some were prostrated and grew long after they fell: these are hard and boxy on one side, hence undesirable. The trees wanted by the miners are those not of the bottom layer, which were *broken down* by the wind or otherwise, and buried at the perfection of their growth.

The first tool of the miner is an iron sounding-rod; with this he probes the mud of the swamp, finding often that the logs lie so thickly across one another beneath the surface that it is only after repeated efforts that he can pass his rod among them. The miner judges of the value of the log he comes in contact with after examination with his probe, by signs known to an expert only; he feels out the size, shape, and position of it, and judges of the work required to secure it; he cuts down to the log through the peat with a sharp spade, and manages to get a chip from it; by *smelling* of this chip he can tell whether he is dealing with a *windfall* or a *breakdown*, the latter being most likely to be sound lumber. Removing the peat, mud, roots, and rubbish-timber as far as necessary, the miner then saws off the log at the ends, his saw working without injury, the soil being free from grit. The log may be thirty feet long, but is generally shorter. Having sawn the log off, the miner uses levers to loosen it from its place and to throw off superincumbent timber; this being done, the log floats upward with perfect buoyance; the under side being *most* buoyant, the log, as it floats free, always turns over. The logs for shingles are sawn into *bolts* or blocks, and *rived* and shaved into shingles on the ground. The ground is gone over again and again with success by the miners, as the logs, once disturbed, continually work toward the surface.

An inch of vegetable matter is deposited by the fall of foliage, twigs, etc., upon the surface of a cedar swamp in about five years, but as this fresh layer is itself buried it partly decays and diminishes in bulk progressively very much by compression and other causes, so that no clue can be had from it as to the age of these remarkable swamps. Such a clue is found, however, in the buried cedars, which by annular rings tell, like a calendar, their own individual age, and by their relative positions demonstrate the successive generations of growth which must have taken place, before they could have appeared where they were left centuries since, superimposed and *grown*, one above another, in many layers.

The attempt to estimate the age chronicled by interwoven logs is confusing, but the certainty of thousands of years is evident, and even ten or twelve generations of such trees as Dr. Beesley examined may have grown and died since the oldest swamp began; and yet the age thus recorded is occupied by the most modern layer of a formation which is, in all and at the oldest, but the very latest evolvement of the most extremely short and insignificant of all the geologic periods.

If the record of ten thousand years can be preserved in mud and perishable wood, what is the chronology of the cycle in which obdurate gneiss and granite grows and disintegrates, crumbles and is recomposed of the old material, again and again, until the Azoic rocks develop into mineral wealth and fertile alluvium, tower into forests, bloom into flowers, ripen into golden harvests, nourish the beasts and birds, redden the blood of the animal world, and give strength and vigor to the body of man; the fitting tabenacle of the immortal soul?

§ Outside the Old beach, and immediately next the strand, are found irregular hillocks of shifting sand of various elevations, but all less than forty feet high. They are composed almost exclusively of fine white quartzose sand without clay or metallic admixture; some few fragments or particles of shell can be found, but the mass of the beach is almost absolutely pure quartz.

Red cedar trees, of which many are dead, are scattered among the hillocks, and are often found buried to the tops where the constantly shifting sands have drifted upon them. These hillocks, downs, or dunes, are denominated *the Little* or *Young Beach;* the method by which they originate is obvious.

Shoal shores at ebb of tide are exposed in wide strands, which rapidly dry under the influence of the sun and wind. Where a wide sandy strand is thus left bare, the wind sweeps the fine sand before it upon the beach and beyond the reach of the returning tide, and then deposits it in the forms described as characteristic of the *Young Beach.* The sea washes up additional sand, which takes the place of that taken off by the wind, and so the process continues which, though counteracted by various agencies, has built up thousands of acres of Young Beach in the State.

The continuity of the beaches on the sea front is broken by a series of inlets, through which the waters of the ocean flow into a number of bays or sounds which lie behind the beaches or within the marshes, and, communicating with one another by inside channels or thoroughfares, make an available still-water navigation for a hundred miles northward from Cape Island to the head of Barnegat Bay. These bodies of water are from five to six miles across in several places, though not on an average more than one-third as wide, having, according to survey and careful estimate, an area of but one hundred and seventeen thousand two hundred and thirty-two acres, excluding the Raritan Bay and adjoining waters.

The most of the inlets themselves are narrow, and, although the tides in the neighborhood of Sea Grove and along shore rise but from a little above four feet in the neap tides to six feet in the spring tides, yet the capacity of the sounds and their adjuncts is so great comparatively that the sea ebbs and flows through the inlets with considerable force, especially when heavy seas run with the incoming tides. In consequence, the bays and sounds are constantly invaded by silt and sand, which, being caught by the abundant growth of grassy marsh roots along their margins, is retained and consolidated in quantities and to a degree which has much decreased the area and depth of those remarkable waters.

The outgoing tide of the sounds cuts away the banks of the inlet and the adjoining shores, throwing the sand out upon the bars, from whence the shore currents and waves convey it along the strand; in this way shoals are added to the southwest of the beaches which crowd the inlets to the south, against the northeast and highest ends of the beaches; presently a new inlet forces its way during a storm across the beach to the northwest of the old one, which may be closed up at the same time. The new inlet is then subjected to the same action as the other one, and with like results; in this way the inlet is continually shifted, wearing its way to the southward for a mile or more at an uncertain rate, and then forcing its way as may be back again to its extreme position toward the north.

Northeast of Barnegat, the inlets move in an opposite manner to the one described as peculiar to those south of that place. The movement of the sands along the New Jersey ocean shore is immense, and due to causes operating on a vast scale in prolonged time. These causes are not fully understood, nor is the scope of their operation fully determined. The theory of the subsidence itself—which, conceding a depression of one-fourth of an inch per annum, would submerge half or more of Cape May County in five hundred and twenty-eight years—has, notwithstanding the facts présumed to demonstrate it, been strenuously disputed by official geologists.

However confident of a conclusion we may feel to be on the basis of facts in our possession, true courage of opinion is not obstinate, and a partial suspense of judgment leaves room for hospitality to the result of enlarged observation, maturer experience, and more deliberate comparison and reflection. Galileo was certain the world moved, and it is equally certain the sands shift. They are carried away from one point or another it may be, but are deposited as well in another place; shoals are created thus, and currents changed, changes of current bring change of drift, and so the transported sands may be shifted back again. Only long-continued observation can establish the fact of a persistent tendency, and still more care must be taken to verify the rate and extent of a movement, the evolution of which is completed only in centuries and ages of time.

§ Beside the quartzose sand composing the beaches along the ocean front of New Jersey, there are an abundance of various kinds of pebbles washed up from the sea. They are of small size, and every year a great quantity of them are conveyed to Philadelphia, and used in roofing buildings and for other purposes. Quartz is exceedingly abundant in South Jersey; the young beach is pure quartz sand, the old beach has but a very small percentage of clay, and the plains or barrens of Burlington and Ocean Counties have from ninety to ninety-eight per cent. of quartz in their soil. Pebbles of pure, transparent quartz abound in the gravel beds and at the shore of Sea Grove and Cape May; being washed clean and bright in the waves, they are often collected by bathing parties and other visitors, under the name of *Cape May Diamonds.* If the collectors have not gained great wealth in their gems, they have often found the treasure of health in their recreative amusements, and some of the pebbles, when well polished and set in gold, form handsome mementos of pleasant summer excursions and peaceful days beside the surf upon the sandy shore, as well as characteristic specimens of a remarkable geologic formation.

Geologic research has, in caves, in mines, in tunnels, and other engineering excavations of submarine nature, been conducted under the sea with interesting and important results, but the present study may end properly and appropriately upon the strand of the open ocean, the type of the unfathomable and infinite. Divesting ourselves of fear, of conceit, of prejudice, of pride; alone with the sand, the waves, the wind, and the breeze—the simplicity and grandeur of nature, our eyes become more clear, the reasoning soul sweeping with one flashing and intuitive glance the old levels and horizons, sees over sea and land a light not born of either, yet luminous as heaven, "which lighteth every man that cometh into the world," revealing even in geologic ruins the love and glory of Our Father, and lighting the way to peace, righteousness, progress, and eternal happiness!

FINIS.

www.ingramcontent.com/pod-product-compliance
Lightning Source LLC
LaVergne TN
LVHW021412110826
845150LV00007B/1889